Electricity
Experiments
You Can Do at Home

About the Author

Stan Gibilisco is an electronics engineer, researcher, and mathematician who has authored *Teach Yourself Electricity and Electronics*, *Electricity Demystified*, more than 30 other books, and dozens of magazine articles. His work has been published in several languages.

The McGraw·Hill Companies

Library of Congress Cataloging-in-Publication Data

Gibilisco, Stan.
    Electricity experiments you can do at home / Stan Gibilisco.
        p.   cm.
    Includes bibliographical references and index.
    ISBN 978-0-07-162164-9 (alk. paper)
    1. Electricity—Experiments.   2. Electric apparatus and appliances—
Experiments.   I. Title.
    TK9901.G487   2010
    537.078—dc22   2010001953

1   2   3   4   5   6   7   8   9   0   DOC/DOC   1   6   5   4   3   2   1   0

ISBN 978-0-07-162164-9
MHID 0-07-162164-4

Sponsoring Editor	**Copy Editor**
Judy Bass	Bhavna Gupta, Glyph International
Editing Supervisor	**Proofreader**
Stephen M. Smith	Priyanka Sinha, Glyph International
Production Supervisor	**Art Director, Cover**
Pamela A. Pelton	Jeff Weeks
Acquisitions Coordinator	**Composition**
Michael Mulcahy	Glyph International
Project Manager	
Vasundhara Sawhney, Glyph International	

Printed and bound by RR Donnelley.

McGraw-Hill books are available at special quantity discounts to use as premiums and sales promotions, or for use in corporate training programs. To contact a representative, please e-mail us at bulksales@mcgraw-hill.com.

This book is printed on acid-free paper.

Electricity Experiments You Can Do at Home

Stan Gibilisco

New York   Chicago   San Francisco   Lisbon   London   Madrid
Mexico City   Milan   New Delhi   San Juan   Seoul
Singapore   Sydney   Toronto

To my physics advisor
at the University of Minnesota
circa 1974
who said:

One experimentalist can keep a dozen theorists busy.

Contents

Preface

This book will educate you, give you ideas, and provoke your curiosity. The experiments described here can serve as a "hands-on" supplement for any basic text on electricity. I designed these experiments for serious students and hobbyists. If you don't have any prior experience with electrical circuits or components, I recommend that you read *Electricity Demystified* before you start here. If you want a deeper theoretical treatment of the subject, you can also read *Teach Yourself Electricity and Electronics*.

If you like to seek out mysteries in everyday things, then you'll have fun with the experiments described in this book. Pure theory might seem tame, but the real world is wild! Some of these experiments will work out differently than you expect. Some, if not most, of your results will differ from mine. In a few scenarios, the results will likely surprise you as they surprised me. In a couple of cases, I could not at the time—and still cannot—explain why certain phenomena occurred.

As I compiled this book, I tried to use inexpensive, easy-to-find parts. I visited the local Radio Shack store many times, browsing their drawers full of components. Radio Shack maintains a Web site from which you can order items that you don't see at their retail outlets. In the back of this book, you'll find a list of alternative parts suppliers. Amateur radio clubs periodically hold gatherings or host conventions at which you can find exotic electrical and electronic components.

As you conduct the experiments described in this book, you're bound to have questions such as "Why are my results so *vastly* different from yours?" If this book stirs up sufficient curiosity and enthusiasm, maybe I'll set up a blog where we can discuss these experiments (and other ideas, too). You can go to my Web site at www.sciencewriter.net or enter my name as a phrase in your favorite search engine.

Have fun!

Stan Gibilisco

Part 1
Direct Current

Your Direct-Current Lab

Every experimenter needs a good workbench. Mine is rather fancy: a piece of plywood, weighted down over the keyboard of an old piano, and hung from the cellar ceiling by brass-plated chains! Yours doesn't have to be that exotic, and you can put it anywhere as long as it won't shake or collapse. The surface should consist of a nonconducting material such as wood, protected by a plastic mat or a small piece of closely cropped carpet (a doormat is ideal). A desk lamp, preferably the "high-intensity" type with an adjustable arm, completes the arrangement.

Protect Your Eyes!

Buy a good pair of safety glasses at your local hardware store. Wear the glasses at all times while doing any experiment described in this book. Get into the habit of wearing the safety glasses whether you think you need them or not. You never know when a little piece of wire will go flying when you snip it off with a diagonal cutters!

Table DC1-1 lists the items you'll need for the experiments in this section. Many of these components can be found at Radio Shack retail stores or ordered through the Radio Shack Web site. A few of them are available at hardware stores, department stores, and grocery stores. If you can't get a particular component from sources local to your area, you can get it (or its equivalent) from one of the mail-order sources listed at the back of this book.

A Bed of Nails

For some of the experiments described in this section, you'll need a prototype-testing circuit board called a *breadboard*. I patronized a local lumber yard to get the wood for the breadboard. I found a length of "12-in by $^3/_4$-in" pine in their scrap heap. The actual width of a "12-in" board is about 10.8 in or 27.4 cm, and the actual thickness is about 0.6 in or 15 mm. They didn't charge me anything for

Table DC1-1 Components list for DC experiments. You can find these items at retail stores near most locations in the United States. Abbreviations: in = inches, AWG = American wire gauge, V = volts, tsp = teaspoon, tbsp = tablespoon, fl oz = fluid ounces, W = watts, A = amperes, K = kilohms, and PIV = peak inverse volts.

Quantity	Store Type or Radio Shack Part Number	Description
1	Lumber yard	Pine board, approx. 10.8 in × 12.5 in × 0.6 in
1	Hardware store	Pair of safety glasses
1	Hardware store	Small hammer
12	Hardware store	Flat-head wood screws, 6 × 32 × $^3/_4$
100	Hardware store	Polished steel finishing nails, $1^1/_4$ in long
1	Department store	12-in plastic or wooden ruler
1	Department store or hardware store	36-in wooden measuring stick (also called a "yardstick")
1	Hardware store	Tube of waterproof "airplane glue" or strong contact cement
1	Hardware store or Radio Shack	Digital multimeter, GB Instruments GDT-11 or equivalent
1	Hardware store	Diagonal wire cutter/stripper
1	Hardware store	Small needle-nose pliers
1	Hardware store	Roll of AWG No. 24 solid bare copper wire
1	278-1221	Three-roll package of AWG No. 22 hookup wire
1	278-1345	Three-roll package of enamel-coated magnet wire
1	Hardware store	Small sheet of fine sandpaper
2	278-1156	Packages of insulated test/jumper leads
1	Hardware store	Heavy-duty lantern battery rated at 6 V
6	Hardware store	Alkaine AA cells rated at 1.5 V
2	270-401A	Holder for one size AA cell
1	270-391A	Holder for four size AA cells in series
1	Grocery store	Pair of thick rubber gloves
1	Grocery store	Small pad of steel wool
1	Hardware store	Galvanized clamping strap, $^5/_8$ in wide
2	Hardware store	Copper clamping strap, $^1/_2$ in wide
1	Grocery store	Container of table salt (sodium chloride)
1	Grocery store	Container of baking soda (sodium bicarbonate)
1	Grocery store	Quart of white distilled vinegar
1	Grocery store	Set of measuring spoons from $^1/_4$ tsp to 1 tbsp
1	Grocery store	Glass measuring cup that can hold 12 fl oz
1	271-1111	Package of five resistors rated at 220 ohms and $^1/_2$ W
1	271-1113	Package of five resistors rated at 330 ohms and $^1/_2$ W
1	271-1115	Package of five resistors rated at 470 ohms and $^1/_2$ W
1	271-1117	Package of five resistors rated at 680 ohms and $^1/_2$ W
1	271-1118	Package of five resistors rated at 1 K and $^1/_2$ W
1	271-1120	Package of five resistors rated at 1.5 K and $^1/_2$ W

Table DC1-1   Components list for DC experiments. You can find these items at retail stores near most locations in the United States. Abbreviations: in = inches, AWG = American wire gauge, V = volts, tsp = teaspoon, tbsp = tablespoon, fl oz = fluid ounces, W = watts, A = amperes, K = kilohms, and PIV = peak inverse volts. (*Continued*)

Quantity	Store Type or Radio Shack Part Number	Description
1	271-1122	Package of five resistors rated at 3.3 K and $^1/_2$ W
1	276-1104	Package of two rectifier diodes rated at 1 A and 600 PIV
1	Department store	Magnetic compass with degree scale, WalMart FC455W or equivalent
1	Department store or office supply store	Small hand-held paper punch that creates $^1/_4$-in holes
2	272-357	Miniature screw-base lamp holder
1	272-1130	Package of two screw-base miniature lamps rated at 6.3 V
1	272-1133	Package of two screw-base miniature lamps rated at 7.5 V
1	277-1205	Encapsulated solar module rated at 6 V output in bright sunlight

the wood itself, but they demanded a couple of dollars to make a clean cut so I could have a fine rectangular piece of pine measuring 12.5 in (31.8 cm) long.

Using a ruler, divide the breadboard lengthwise at 1-in (25.4-mm) intervals, centered so as to get 11 evenly spaced marks. Do the same going sideways to obtain 9 marks at 1-in (25.4-mm) intervals. Using a ball-point or roller-point pen, draw lines parallel to the edges of the board to obtain a grid pattern. Label the grid lines from A to K and 1 to 9 as shown in Fig. DC1-1. That'll give you 99 intersection points, each of which can be designated by a letter-number pair such as D-3 or G-8.

Once you've marked the grid lines, gather together a bunch of 1.25-in (31.8-mm) polished steel finishing nails. Place the board on a solid surface that can't be damaged by scratching or scraping. A concrete or asphalt driveway is ideal for this purpose. Pound a nail into each grid intersection point as shown in Fig. DC1-1. Be sure the nails are made of polished steel, preferably with "tiny heads." The nails must not be coated with paint, plastic, or any other insulating material. Each nail should go into the board just far enough so that you can't wiggle it around. I pounded every nail down to a depth of approximately 0.3 in (8 mm), halfway through the board.

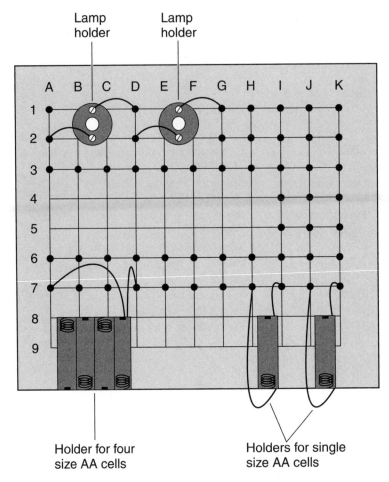

Figure DC1-1 *Layout of the breadboard for DC experiments. I used a "12-in" pine board (actually 10.8 in wide) with a thickness "3/4 in" (actually about 0.6 in), cut to a length of 12.5 in. Solid dots show the positions of the nails. Grid squares measure 1 in by 1 in.*

Lamp and Cell Holders

Using 6 × 32 flat-head wood screws, secure the two miniature lamp holders to the board at the locations shown in Fig. DC1-1. Using short lengths of thin, solid, bare copper wire, connect the terminals of one lamp holder to breadboard nails A-2 and D-1. Connect the terminals of the other lamp holder to nails D-2 and G-1. Wrap the wire tightly at least twice, but preferably four times, around each nail. Snip off any excess wire that remains.

Glue two single-cell AA battery holders and one four-cell AA battery holder to the breadboard with contact cement. Allow the cement to harden for 48 hours. Then strip 1 in of the insulation from the ends of the cell-holder leads and connect the leads to the nails as shown in Fig. DC1-1. Remember that the red leads are positive and the black leads are negative. Use the same wire-wrapping technique that you used for the lamp-holder wires. Place fresh AA cells in the holders with the negative sides against the springs. Your breadboard is now ready to use.

Wire Wrapping

The breadboard-based experiments in this book employ a construction method called *wire wrapping*. Each of the nails in your breadboard forms a terminal to which several component leads or wires can be attached. To make a connection, wrap an uninsulated wire or lead around a nail in a tight, helical coil. Make at least two, but preferably four or five, complete wire turns as shown in Fig. DC1-2.

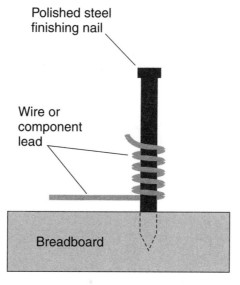

Polished steel
finishing nail

Wire or
component
lead

Breadboard

Figure DC1-2   *Wire-wrapping technique. Wind the wire or component lead at least twice, but preferably four or five times, around the nail. Extra wire should be snipped off if necessary, using a diagonal cutter.*

When you wrap the end of a length of wire, cut off the excess wire after wrapping. For small components such as resistors and diodes, wrap the leads around the nails as many times as is necessary to use up the entire lead length. That way, you won't have to cut down the component leads. You'll be able to easily unwrap and reuse the components for later experiments. Needle-nose pliers can help you to wrap wires or leads that you can't wrap with your fingers alone.

When you want to make multiple connections to a single nail, you can wrap one wire or lead over the other, but you shouldn't have to do that unless you've run out of nail space. Each nail should protrude approximately 1 in above the board surface, so you won't be cramped for wrapping space. Again, let me emphasize that the nails should be made of polished steel without any coating. They should be new and clean, so they'll function as efficient electrical terminals.

Let's Get Started!

When you perform the experiments in this section, the exact arrangement of parts on the breadboard is up to you. I've provided schematic and/or pictorial diagrams to show you how the components are interconnected.

Small components such as resistors and diodes should be placed between adjacent nails, so that you can wrap each lead securely around each nail. *Jumper wires* (also known as *clip leads*) should be secured to the nails so that the "jaws" can't easily be pulled loose. It's best to clamp jumpers to nails sideways, so that the wires come off horizontally.

Caution!   *Use needle-nose pliers and rubber gloves for any wire-wrapping operations if the voltage at any exposed point might exceed 10 V.*

Caution!   *Wear safety glasses at all times as you do these experiments, whether you think you need the glasses or not.*

DC2

Voltage Sources in Series

In this experiment, you'll find out what happens when you connect electrical cells or batteries in *series* (that is, end-to-end) in the same direction. Then you'll discover what occurs when you connect one of the cells in the wrong direction. Finally, you'll get a chance to do your own experiment and see if you can predict what will take place.

What's a Volt?

Current can flow through a device or system only if electrical *charge carriers* (such as electrons) are "pushed" or "motivated." The "motivation" can be provided by a buildup of charge carriers, with *positive polarity* (a shortage of electrons) in one place and *negative polarity* (an excess of electrons) in another place. In a situation like this, we say that an *electromotive force* (EMF) exists. This force is commonly called *voltage*, and it's expressed in units called *volts* (symbolized V). You'll occasionally hear voltage spoken of as *electrical potential* or *potential difference*.

How large is a potential difference of 1 V? You can get an idea when you realize that a flashlight cell produces about 1.5 V, a lantern battery about 6 V, an automotive battery 12 to 14 V, and a standard household utility outlet 110 to 120 V. Cells and batteries produce *direct-current* (DC) voltages, while household utility systems in the United States produce *alternating-current* (AC) voltages.

For the following set of experiments, you'll need two size AA flashlight cells rated at 1.5 V, one lantern battery rated at 6 V, and a digital meter capable of measuring low DC voltages, accurate to within 0.01 V.

Cell and Battery Working Together

In an *open circuit* where series-connected cells or batteries aren't hooked up to any external device or system, the voltages always add up, as long as we connect the cells in the same direction. This is true even if the individual cells are of different electrochemical types, such as zinc-carbon, alkaline, or lithium-ion. For this rule to hold, however, you must be sure that all the cells are connected plus-to-minus, so that they work together. The rule does not apply if any of the cells is reversed so that its voltage *bucks* (works against, rather than with) the voltages produced by the other cells.

When I measured the voltages of the new AA cells I purchased for this experiment, each cell tested at 1.58 V. The lantern battery produced 6.50 V. Your cells and battery will probably have different voltages than mine did, so be sure that you test each cell or battery individually.

According to basic electricity theory, I expected to get 8.08 V when I connected one AA cell in series with the lantern battery and then measured the voltage across the whole combination. By simple addition,

$$1.58 \text{ V} + 6.50 \text{ V} = 8.08 \text{ V}$$

There were two different ways to connect these two units in series. Figure DC2-1 shows the arrangements, along with the theoretical and measured voltages. These wiring diagrams use the standard schematic symbols for a cell, a battery, and a meter. The voltages I measured for the series combinations were both 10 *millivolts* (mV) less than the theoretical predictions. A millivolt is equal to 0.001 V, so 10 mV is 0.01 V. That's not enough error to be of any concern. Small discrepancies like this are common in physical science experiments.

To build the battery-cell combination, I placed the battery terminal against the cell terminal, holding the two units together while manipulating the meter probes. I grasped the probes and the cell to keep the arrangement from falling apart. The voltages in this experiment weren't dangerous, but I wore rubber gloves (the sort people use for washing dishes) to keep my *body resistance* from affecting the voltage readings.

Cell Conflicting with Battery

In the arrangement shown by Fig. DC2-1, we actually have five electrochemical cells connected in series, because the lantern battery contains four internal cells. If you reverse the polarity of the AA cell, you might be tempted to suppose that its voltage will subtract from that of the battery, instead of adding to it. However, in

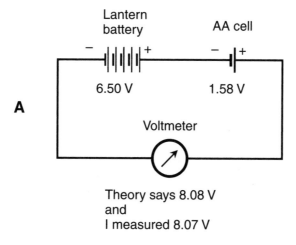

A

Lantern battery

AA cell

6.50 V

1.58 V

Voltmeter

Theory says 8.08 V
and
I measured 8.07 V

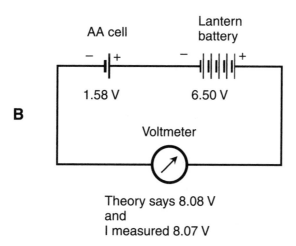

B

AA cell

Lantern battery

1.58 V

6.50 V

Voltmeter

Theory says 8.08 V
and
I measured 8.07 V

Figure DC2-1   *Here's what happened when I connected a lantern battery and a flashlight cell in series so that they worked together. At A, the positive pole of the battery went to the negative pole of the cell. At B, the positive pole of the cell went to the negative pole of the battery.*

physical science, suppositions and assumptions often turn out to be wrong! Let's test such an arrangement and see what really occurs.

Figure DC2-2 shows two ways to connect the AA cell so that its voltage bucks the battery voltage. Theory predicts 4.92 V across either of these series combinations, because

$$6.50 \text{ V} - 1.58 \text{ V} = 4.92 \text{ V}$$

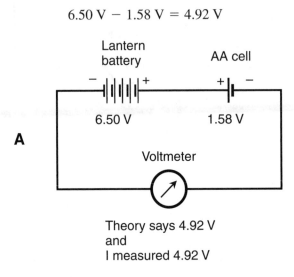

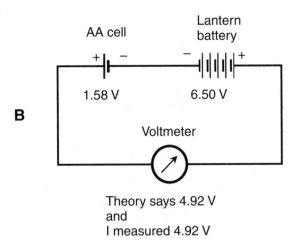

Figure DC2-2   *When I connected a lantern battery and flashlight cell in series so that they worked against each other, the cell's voltage took away from the battery's voltage. At A, plus-to-plus; at B, minus-to-minus.*

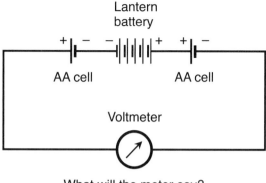

What will the meter say?

Figure DC2-3   *What will happen when two flashlight cells are connected in series with a battery as shown here, so that both cells fight against the battery and a voltmeter is connected across the entire combination?*

When I did the experiment, I got exactly this result. Either way I connected the two voltage sources to buck each other—plus-to-plus (as in Fig. DC2-2A) or minus-to-minus (as in Fig. DC2-2B)—the net output voltage across the combination was the same. In this case, theory and practice agreed. In Experiment DC3 you'll see a situation where they don't.

Now Try This!

Once again, measure the voltages of the two AA cells and the lantern battery all by themselves. Write them down. Connect the cells to the battery as shown in Fig. DC2-3, with one cell on each pole of the battery. One connection should be plus-to-plus, and the other connection should be minus-to-minus. What do you think the meter will say, according to electricity theory, when it's connected across the whole combination? Make the measurement, and see if you're right.

DC3

Current Sources in Series

You've seen that the open-circuit voltages of flashlight cells add and subtract when you connect them in series. Now you'll find out how much current such a combination can deliver. You'll need four size AA cells, a cell holder for four AA cells in series, and a DC ammeter that can measure at least 10 amperes (also called *amps*).

What's an Ampere?

Theoretically, an electric current is measured in terms of the number of charge carriers, usually electrons, that pass a point in 1 second. But in practice, current is rarely expressed directly in that manner. Instead, engineers express current in units of *coulombs per second*, where 1 coulomb is approximately 6,240,000,000,000,000,000 (6.24 quintillion). This quantity can be written in *scientific notation* as 6.24×10^{18}. One coulomb per second represents an *ampere* (symbolized A), the standard unit of electric current.

Maximum Deliverable Current

In theory, an ideal voltmeter doesn't draw any current at all. In practice, a good voltmeter is designed to draw as little current as possible from circuits under test. It therefore has an extremely high *resistance* between its terminals. When you measured the voltages in the last chapter, you weren't making the cells or the battery "do any work." In this experiment, you'll measure the *maximum deliverable current* values for small cells and combinations. You'll use an ammeter instead of a voltmeter.

A theoretically ideal ammeter would be a perfect *short circuit*. A real-world ammeter is designed to have an extremely low resistance between its terminals. When you connect an ammeter directly across a cell or battery, you "short out" the cell or battery, forcing it to produce all the current that it can. The maximum deliverable current is limited by the internal resistance of the ammeter, the internal resistance of the cell or battery, and the resistance of the circuit wiring.

Caution!   *This experiment involves short-circuiting electrochemical cells and batteries. Never leave the ammeter terminals connected for more than 2 seconds at a time. After making a measurement of the maximum deliverable current, wait at least 10 seconds before connecting the ammeter to the cell or battery again. Longer connections can cause cells or batteries to overheat, leak, or rupture. Don't use cells larger than AA size.*

One Cell Alone

Begin by measuring the maximum deliverable currents for each cell individually. Call the cells #1, #2, #3, and #4. (It's a good idea to write the numbers on the cells with an indelible marker to keep track of which one is which.) I used four new alkaline cells, fresh out of the package. Each cell produced approximately 9.25 A when shorted out by the ammeter, as shown in Fig. DC3-1A. The digital reading fluctuated because of the inexact nature of current measurements using ordinary wires and probes, so I had to estimate it.

Four Cells Working Together

Place the cells in the holder so that their voltages add up. The negative terminals should rest against the spring contacts, while the positive terminals rest against the flat contacts. My cell holder has two wires coming out of it, both stripped at the ends. Black is negative; red is positive.

When I held the ammeter probes firmly against the exposed metal ends of the wires, making sure that the meter was set to handle 10 A or more, I got a reading of approximately 9.25 A (Fig. DC3-1B). Again, the reading fluctuated. The digits wouldn't "settle down," so I had to make a visual estimate.

The voltage produced by the series cell combination is four times the voltage from a single cell. It's reasonable to suppose that the total internal battery resistance is also four times as great as that of a single cell. *Ohm's law* tells us that current equals voltage divided by resistance, so it's no surprise that the maximum deliverable current of the series combination is the same as that of any cell by

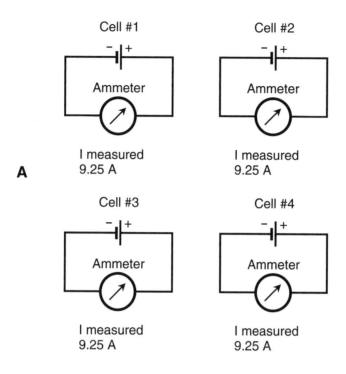

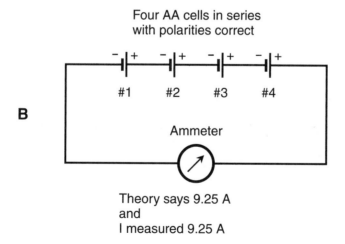

Figure DC3-1   *I short-circuited four different AA cells with an ammeter, and estimated that each cell produced 9.25 A as shown at A. Then I connected the cells in series so that they all worked together, shorted the combination with the ammeter, and got 9.25 A again as shown at B.*

itself. We've increased the voltage by a factor of 4, but we've also increased the internal resistance by a factor of 4, and $4/4 = 1$.

One Cell Conflicting

What do you think will happen if you reverse the polarity of one of the four cells in the arrangement of Fig. DC3-1B and then measure the maximum deliverable current of the combination? You already know that if you reverse one of the cells in an open circuit, then its voltage subtracts from the total instead of adding to it, halving the voltage of four identical series-connected cells. If the total internal resistance of the combination stays the same, then Ohm's law suggests that the current in either of the arrangements in Fig. DC3-2 should be half the current in the arrangement of Fig. DC3-1B, or something like 4.6 or 4.7 A.

When I did the experiment, I didn't get currents anywhere near these theoretical predictions! Instead, I got 6 A or a little more. As if that wasn't strange enough, the current in the arrangement of Fig. DC3-2A was slightly different from the current in the arrangement of Fig. DC3-2B. The theory that I suggested in the previous paragraph was proven invalid by my own experiment. I had to conclude that the internal resistance of the cell combination changed when one of the cells was turned around. But if that was true, why did reversing cell #1 change the internal resistance to a different extent than reversing cell #4? I can't explain it. Can you?

Now Try This!

In case you discover a theory that accurately predicts the results of the tests shown in Fig. DC3-2, here's another experiment you can try. Connect four AA cells in series with all the polarities correct (that is, so that they all work together), and then reverse one of the cells in the middle. This can be done in two ways, as shown in Fig. DC3-3. Measure the maximum deliverable current under these conditions. What do you think will happen? What takes place when you actually test the circuits?

Variations

You ought to have figured out by now that this experiment is inexact by nature. Your results will probably differ from mine, depending on the ages and chemical compositions of your cells. If you like, try different types of cells, such as

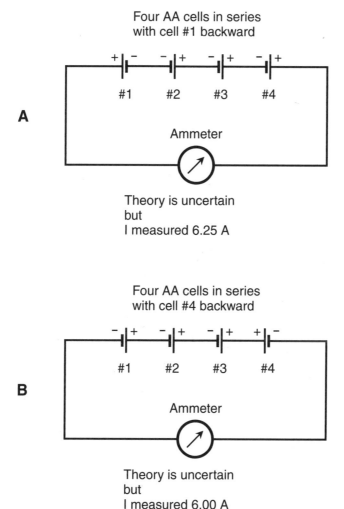

A

Four AA cells in series
with cell #1 backward

#1   #2   #3   #4

Ammeter

Theory is uncertain
but
I measured 6.25 A

B

Four AA cells in series
with cell #4 backward

#1   #2   #3   #4

Ammeter

Theory is uncertain
but
I measured 6.00 A

Figure DC3-2   *When I connected one of the end cells backward, I expected to get 4.6 or 4.7 A, but in real life the current was higher. With the arrangement shown at A, I measured about 6.25 A; with the arrangement shown at B, I measured about 6.00 A.*

nickel-metal-hydride or lithium-ion. But again, don't use cells larger than AA size, and don't "short them out" for more than 2 seconds at a time. How do the maximum deliverable currents compare for different cell ages and types? Can you invent a theory that accurately predicts your test results in all cases?

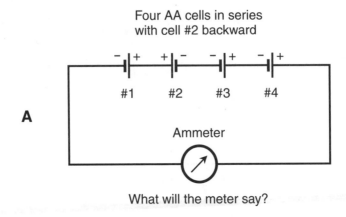

Four AA cells in series
with cell #2 backward

A

What will the meter say?

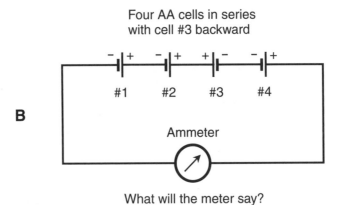

Four AA cells in series
with cell #3 backward

B

What will the meter say?

Figure DC3-3   *What do you think will be the maximum deliverable current of a four-cell series combination with the second cell reversed as shown at A, or with the third cell reversed as shown at B?*

A Simple Wet Cell

In this experiment, you'll build an electrochemical *wet cell* and see how much voltage and current it can produce. You'll need a short, fat, thick glass cup that can hold 12 ounces (oz) (about 0.36 liter [L]) when full to the brim. You'll need some distilled white vinegar and two pipe clamps measuring $1/_2$ to $5/_8$ inch (in) (1.3 to 1.6 centimeters [cm]) wide, one made of copper and the other of galvanized steel, designed to fit pipes 1 in (2.5 cm) in diameter. You'll also need some bell wire.

Setting It Up

Get rid of the bends in the pipe clamps, and straighten them out into strips. The original clamps should be large enough so that the flattened-out strips measure at least 4 in (about 10 cm) long. Polish both sides of the strips with steel wool or a fine emery cloth to get rid of any layer of *oxidation* that might have formed on the metal surfaces.

Strip 2 in (5 cm) of insulation from each end of two 18-in lengths of bell wire. Attach a length of bell wire to each electrode by passing one stripped end of the wire through one of the holes in the electrode and wrapping the wire around two or three times as shown in Fig. DC4-1A. Wrap the "non-electrode" end of the stripped wire from the copper electrode around the positive (red) meter probe tip as shown in Fig. DC4-1B. Wrap the "non-electrode" end of the wire from the galvanized electrode around the negative (black) meter probe tip in the same way. Secure all connections with electrical tape to insulate them and keep them stable. Remove both of the meter probe leads from their receptacles on the meter.

Lay the strips against the inside sides of the cup with their ends resting on the bottom. Be sure that the strips are on opposite sides of the cup, so they're as far away from each other as possible. Bend the strips over the edges of the cup to hold them in place, as shown in Fig. DC4-2. Be careful not to break the glass! Fill the cup with vinegar until the liquid surface is slightly below the brim.

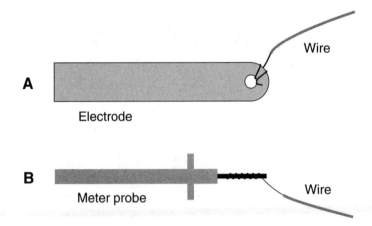

Figure DC4-1   *Attachment of wires to the electrodes (at A) and the meter probes (at B). Wrap the bare wire around the metal. Then secure the connections with electrical tape.*

Add Salt

Once you've put the parts together as shown in Figs. DC4-1 and DC4-2, add one rounded teaspoon of common table salt (sodium chloride). Stir the mixture until the salt is completely dissolved in the vinegar. You'll know that all the salt has

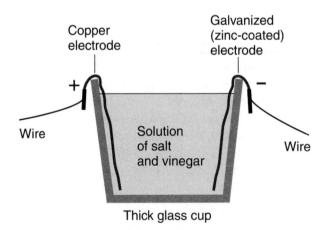

Figure DC4-2   *A wet cell made from a vinegar-and-salt solution. The glass cup has a brimful capacity of approximately 12 fluid oz (0.36 L).*

dissolved when you don't see any salt crystals on the bottom of the cup after you allow the liquid to stand still for a minute.

Set the meter to measure a low DC voltage. The best meter switch position is the one that indicates the smallest voltage that's greater than 1 volt (V). Insert the negative meter probe lead into its receptacle on the meter. Then insert the positive meter probe lead and note the voltage on the meter display. When I conducted this experiment, I got a reading of 515 millivolts (mV) (or 0.515 V). After 60 seconds, the voltage was still 515 mV.

Remove the positive meter lead from its receptacle on the meter. Set the meter for a low DC current range. The ideal setting is the lowest one showing a maximum current of 20 *milliamperes* or more. A milliampere (also called a *milliamp* and symbolized mA) equals 0.001 ampere (A). Insert the disconnected meter lead back into its receptacle, and carefully note how the current varies with time. I got a reading of 8.30 mA to begin with. The current dropped rapidly at first, then more and more slowly. After 60 seconds, the current stabilized at 7.45 mA, as shown by the lowermost (solid) curve in Fig. DC4-3.

When you conduct these tests, you'll probably get more or less voltage or current than I got, depending on how much vinegar is in your cup, how strong the vinegar is, and how large your electrodes are. In any case, you should find that the open-circuit voltage remains constant as time passes, while the maximum deliverable current decreases.

Caution!   *In this experiment, you don't have to worry about "shorting out" the cell for more than 2 seconds. The cell doesn't produce anywhere near enough energy to boil the vinegar-and-salt* electrolyte, *and the electrolyte can't leak because it's in the open to begin with. But if you get a notion to try any of these exercises with an automotive battery or other large commercial wet cell or battery,* forget about it! *The electrolyte in that type of device is a powerful and dangerous acid that can violently boil out if you short-circuit the terminals.*

Add More Salt

Add another rounded teaspoon of salt to the vinegar. As before, stir the solution until the salt has completely dissolved. Repeat the voltage and current experiments. You should observe slightly higher voltages and currents. As before, the open-circuit voltage should remain constant over time, and the maximum deliverable current should fall. I measured a constant 528 mV. The current started out at 10.19 mA and declined to 8.76 mA after 60 seconds, as shown by the middle (dashed) curve in Fig. DC4-3.

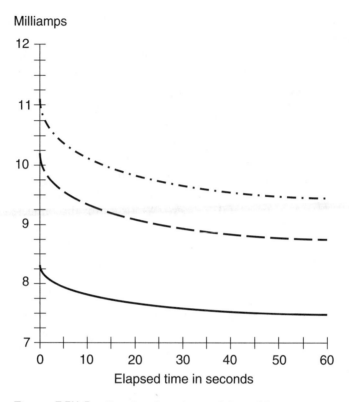

Figure DC4-3   *Graphs of maximum deliverable currents as functions of time for various amounts of salt dissolved in 12 fluid oz (0.36 L) of vinegar. Lower (solid) curve: one rounded teaspoon of salt. Middle (dashed) curve: two rounded teaspoons of salt. Upper (dashed-and-dotted) curve: three rounded teaspoons of salt.*

Add a third rounded teaspoon of salt and fully dissolve it. Once again, measure the open-circuit voltage and the maximum deliverable current. When I did this, I got a constant 540 mV. The current began at 11.13 mA, diminishing to 9.43 mA after 60 seconds passed, as shown by the uppermost (dashed-and-dotted) curve in Fig. DC4-3.

The increased voltage and current with added salt is the result of greater chemical activity of the electrolyte. If you add still more salt beyond the three rounded teaspoons already in solution, you'll eventually reach a point where the vinegar can't take any more. The solution will be *saturated*, and the electrolyte will have reached its greatest possible concentration.

Discharge and Demise

When you measure the voltage across the terminals of your wet cell without requiring that the cell deliver any current (other than the tiny amount required to activate the voltmeter), the cell doesn't have to do any work. You might expect that the voltage will remain constant for hours. Let the cell sit idle overnight, with nothing connected to its terminals, and measure its voltage again tomorrow. What do you think you'll see?

When you have the ammeter connected across the cell terminals, you'll notice that bubbles appear on the electrodes, especially with higher salt concentrations. The bubbles consist of gases (mainly hydrogen and oxygen, but also some chlorine) created as the electrolyte solution breaks down into its constituent elements. Although you won't see it in a short time, the electrodes will become coated with solid material as well.

If you "short out" your wet cell and leave it alone for an extended period of time, all of the chemical energy in the electrolyte will eventually get converted into heat. The maximum deliverable current will fall to zero, as will the open-circuit voltage. The cell will have met its demise.

Now Try This!

Conduct this experiment with different salts, such as potassium chloride (salt substitute) or magnesium sulfate (also known as Epsom salt). Then try it with lemon juice instead of vinegar. How do the results vary? Plot the open-circuit voltages and maximum deliverable currents graphically as functions of time, and compare these graphs with the curves in Fig. DC4-3.

DC5

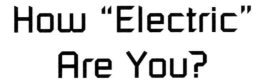

How "Electric" Are You?

A wet cell works because of chemical reactions between dissimilar metal electrodes and an electrolyte solution. In Experiment DC4, you used vinegar and salt as the electrolyte. In this experiment, you'll use the same solution to make contact between the electrodes and your hands, but most of the electrolyte will be your own "flesh and blood"! For this experiment, you'll need all the items left over from Experiment DC4.

Setting It Up

Remove the galvanized and copper electrodes from the vinegar-and-salt solution. Leave the solution in the cup. Leave the wires connected to the electrodes. Rinse the electrodes with water, dry them off, and get rid of the bends so they're both flat strips with holes in each end. Make sure that the probe leads are plugged into the meter. Then switch the meter to one of the more sensitive DC voltage ranges.

Body Voltage

Wet your thumbs, index fingers, and middle fingers up to the first knuckles by sticking both hands into the vinegar-and-salt solution. (Don't be surprised if this solution stings your fingers a little bit. It's harmless!) Grasp the electrodes between your thumb and two fingers. Don't let your hands come into contact with the wires, but only with the metal faces of the electrodes. What does the meter say? When I conducted this experiment, I got a steady voltage of 515 millivolts (mV).

Body Current

Rinse your hands with water and dry them off. Switch the meter to the most sensitive DC current range. In my meter, that's a range of 0 to 200 *microamperes*. (A micro-ampere, also called a *microamp* and symbolized μA, is 0.001 mA or 0.000001 A.) Wet your fingers with the vinegar-and-salt solution again, and grasp the electrodes in the same way as you did when you measured the voltage. Watch the current level for 60 seconds, making sure that you don't change the way you hold onto the electrodes. The current reading should decline, rapidly at first, and then more slowly.

When I did this experiment, the current started out at 122 μA and declined to 95 μA after 60 seconds had passed. Beyond 60 seconds, the current remained almost constant. Figure DC5-1 illustrates the current-vs.-time function as a graph.

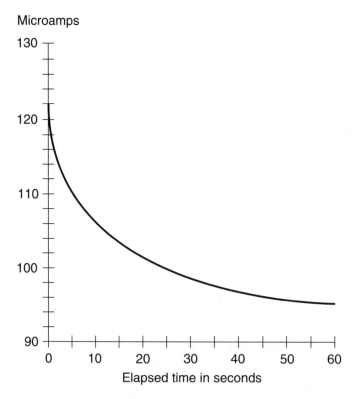

Figure DC5-1   *Graph of maximum deliverable current as a function of time from my "body cell." My hands were wetted with a solution of three rounded teaspoons of salt dissolved in 12 fluid ounces (fl oz) of vinegar.*

Now Try This!

Set the meter to a different current range and repeat the above experiment. You should not expect to get the same readings as before. Of course, a small amount of variation is inevitable in any repeated experiment involving material objects. In this case, however, you should see a difference that's too great to be explained away by imperfections in the physical hardware.

Set the meter back to the same range you used in the first current-measuring experiment, and do it that way again. Then set the meter to the range you used in the second experiment, and go for yet another round. Keep switching back and forth between the two meter ranges, rinsing off your fingers, drying them, and rewetting them with electrolyte solution each time. Do you get more body current with the meter set to measure the higher range (less sensitive) than you do with the meter set to measure the lower range (more sensitive)?

What's Happening?

An ideal ammeter would have no internal resistance, so it would have no effect on the behavior of a circuit when connected in series with that circuit. But in the real world, all ammeters have some internal resistance, because the wire coils inside them don't conduct electricity perfectly. Unless it's specially engineered to exhibit a constant internal resistance, a meter that's set to measure small currents has a greater internal resistance than it does when it's set to measure larger currents. Most inexpensive test meters (such as mine!) aren't engineered to get rid of these little discrepancies.

When I changed the meter range while measuring my body current, my meter's internal resistance competed with my body's internal resistance. When I set the meter to a lower current range, I increased the total resistance in the circuit, reducing the actual flow of current. Conversely, as I set the meter to a higher current range, I decreased the total resistance in the circuit, increasing the actual current.

What Should We Believe?

Does this phenomenon remind you of the *uncertainty principle* that physicists sometimes talk about? The behavior of an observer can change the behavior of the observed system. Sometimes, if not usually, this effect is too small to see. In this experiment, if you have the same type of meter that I have, the uncertainty principle is vividly apparent.

So, you might ask, which of my body current measurements should I believe? The answer: All of them. My current meter might not be an ideal device, but it faithfully reports what it sees: the number of coulombs of electrical charge carriers that pass through it every second.

DC6

Your Body Resistance

In Experiment DC5, you discovered that your body resistance affects the amount of current that can flow in a circuit when you're part of that circuit. In this experiment, you'll measure your body resistance. You'll need everything you used in Experiment DC5, along with a second copper electrode.

What's an Ohm?

The standard unit of resistance is the *ohm*, which engineers and technicians sometimes symbolize with the uppercase Greek letter omega (Ω). The ohm can be defined in two ways:

- The amount of resistance that allows ampere (1 A) of current to flow when volt (1 V) of *electromotive force* is applied across a component or circuit, or
- The amount of resistance that allows 1 V of potential difference to exist across a component or circuit when 1 A of current flows through it.

As resistance increases, conductance decreases. As conductance increases, resistance decreases. The smallest possible resistance is 0 ohms, representing a component or circuit that conducts electricity perfectly. There's no limit to how large a resistance can become; an open circuit is sometimes said to have a resistance of "infinity." When expressing large resistances, you might want to use the *kilohm* (symbolized K), which is equal to 1000 ohms, or the *megohm* (symbolized M), which is equal to 1,000,000 ohms.

How Resistance Is Measured

An *ohmmeter* (resistance-measuring meter) for DC can be constructed by placing a DC milliammeter or microammeter in series with a set of fixed, switchable

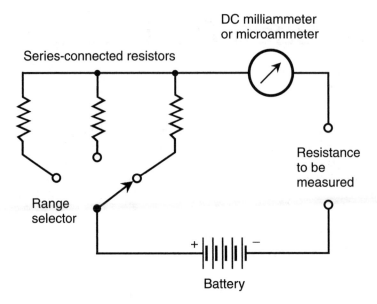

Figure DC6-1 *A multirange ohmmeter works by switching various resistors of known values in series with a sensitive DC current meter.*

resistances and a battery that provides a known, constant DC voltage, as shown in Fig. DC6-1. By selecting the resistances appropriately, the meter gives indications in ohms over any desired range. The device can be set to measure resistances from 0 ohms up to a certain maximum such as 2 ohms, 20 ohms, 200 ohms, 2 K, 20 K, 200 K, 2 M, or 20 M.

An ohmmeter must be calibrated at the factory where it is made, or in an electronics lab. A small error in the values of the series resistors can cause large errors in measured resistance. Therefore, these resistors must have precise *tolerances*. In other words, their values must actually be what the manufacturer claims they are, to within a fraction of 1 percent if possible. In addition, the battery must provide exactly the right voltage.

If you want to measure the resistance between two points with an ohmmeter, you must be sure that no voltage exists between the points where you intend to connect the meter. Such a preexisting voltage will add or subtract from the ohmmeter's internal battery voltage, producing a false reading. Sometimes, in this type of situation, an ohmmeter might say that the component's resistance is less than 0 ohms or more than "infinity"!

How Resistive Are You?

The measurement of internal body resistance is a tricky business. The results you get will depend on how well the electrodes are connected to your body, and also on where you connect them.

Get a second copper clamp from your "junk box." Take the bends out of it so it's a flat strip, and then polish it in the same way as you polished the other two electrodes. Connect one copper strip to each of the meter probe tips using bell wire. Switch the meter to measure a relatively high resistance range, say 0 to 20 K. Dip your fingers into the electrolyte solution left over from Experiment DC5. What does the meter say? Repeat the experiment using the next higher resistance range (in my meter, that would be 0 to 200 K).

When I measured my body resistance using the above-described scheme, I got approximately 7.8 K (that is, 7800 ohms) with the meter set for 0 to 20 K, and 4.9 K with the meter set for 0 to 200 K. The difference resulted from internal meter resistance, just as in Experiment DC5 with the current-measuring apparatus. The higher resistance range required a different series resistance than the lower range. These resistances appeared in series with the resistance of my body, so the total current flow (which is what the meter actually "sees") changed as the range switch position changed.

A friend of mine tried this experiment. He got 6.3 K at the 0-to-20-K meter range, and 4.5 K at the 0-to-200-K meter range. He wondered if the results of this experiment could be an indicator of a person's overall health. I said that I didn't think so, but only his doctor would know for sure.

Now Try This!

Try this experiment with a copper electrode and a galvanized electrode, in the same arrangement as you used when you performed Experiment DC5. Connect your body to the meter as shown in Fig. DC6-2A. Then reverse the polarity of your "body-electrode-meter" circuit by connecting the red wire to the black meter input, and connecting the black wire to the red meter input, so you get the configuration shown in Fig. DC6-2B. You should observe different meter readings. You might even get a "negative" body resistance or a meter indication to the effect that the input is invalid.

Once you've completed this part of the experiment, remember to return the meter probes to their correct positions: black probe to black jack, and red probe to red jack.

Why do you think a discrepancy in the meter readings occurs when the electrode metals are dissimilar, but not when they're identical? Your body resistance

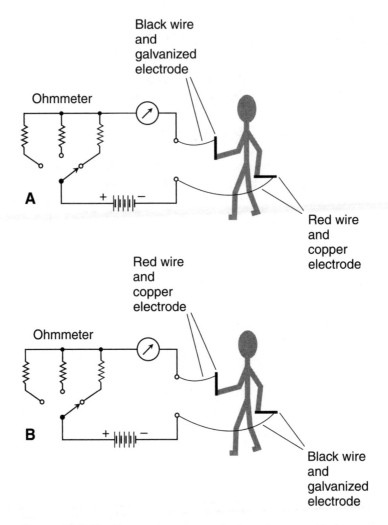

Figure DC6-2 *Try to measure your body resistance with the arrangement you used to measure current in Experiment DC5, as shown at A. Then try the same test with the meter probe wires reversed, as shown at B.*

doesn't depend on the direction in which electrons travel from atom to atom through your blood and bones—does it?

DC7

Resistances of Liquids

In this experiment, you'll measure the resistance of "pure" tap water, and then add two different mineral salts to create solutions that increase the conductivity. You'll need everything from Experiment DC6 except the vinegar. You'll also need a chef's measuring spoon with a capacity of 0.5 teaspoon.

Resistance of Tap Water

Connect both meter probe wires to copper electrodes as you did when you measured your body resistance in Experiment DC6. Plug the negative (black) meter probe jack into the common-ground meter input, but leave the positive (red) probe jack unplugged. Place the two copper electrodes into the cup as you did in Experiment DC4. Be sure that the strips are on opposite sides of the cup, so they're as far away from each other as possible. Bend the strips over the edges of the cup to hold them in place, as shown at A in Fig. DC7-1. Fill the cup with water until the liquid surface is slightly below the brim. Use tap water, not distilled or bottled water.

Switch your ohmmeter to measure resistances in a range from 0 to several kilohms (K). Find a clock or watch with a second hand, or a digital clock or watch that displays seconds as they pass. When the second hand reaches the "top of the minute" or the digital seconds display indicates "00," plug the positive meter probe jack into its receptacle. Note the resistance at that moment. Then, keeping one eye on the clock and the other eye on the ohmmeter display, note and record the resistances at 15-second intervals until 90 seconds have elapsed. When I did this experiment, I got the following results:

- At the beginning: 3.49 K

- After 15 seconds: 3.82 K

- After 30 seconds: 4.10 K

- After 45 seconds: 4.30 K

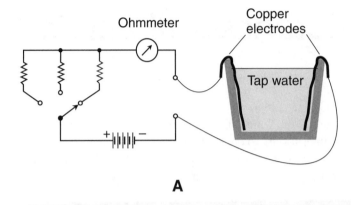

A

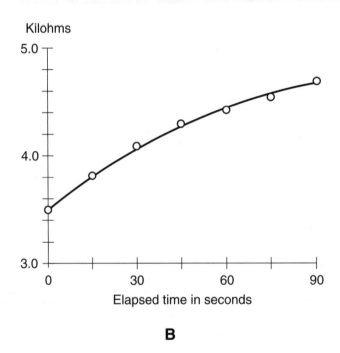

B

Figure DC7-1 *At A, the arrangement for measuring the
resistance of tap water. At B, the resistance values as I measured them over a time span of 90 seconds.*

- After 60 seconds: 4.42 K

- After 75 seconds: 4.54 K

- After 90 seconds: 4.68 K

Figure DC7-1B graphically shows these results. Open circles are plotted data points. The black curve is an optimized graph obtained by *curve fitting*. Your results will differ from mine depending on the type of ohmmeter you use, the dimensions of your electrodes, and the mineral content of your tap water. In any case, you should observe a gradual increase in the resistance of the water as time passes.

Resistance of Salt Water

Remove the positive meter probe jack from its meter receptacle. Switch the ohmmeter to the *next lower* resistance range. Carefully measure out 0.5 teaspoon of table salt (sodium chloride). Use a calibrated chef's cooking spoon for this purpose, and level off the salt to be sure that the amount is as close to 0.5 teaspoon as possible. Pour the salt into the water. Stir the solution until the salt is completely dissolved. Then allow the solution to settle down for a minute.

Once again, watch the clock. At the "top of the minute," plug the positive meter probe jack into its receptacle. Your system should now be interconnected as shown at A in Fig. DC7-2. Record the resistances at 15-second intervals. Here are the results I got:

- At the beginning: 250 ohms

- After 15 seconds: 262 ohms

- After 30 seconds: 269 ohms

- After 45 seconds: 273 ohms

- After 60 seconds: 276 ohms

- After 75 seconds: 282 ohms

- After 90 seconds: 286 ohms

Remove the positive meter probe jack from its outlet. Add another 0.5 teaspoon of salt to the solution, stir it in until it's totally dissolved, and then let the solution settle down. Repeat your timed measurements. I got the following results:

- At the beginning: 165 ohms

- After 15 seconds: 193 ohms

- After 30 seconds: 201 ohms

- After 45 seconds: 207 ohms

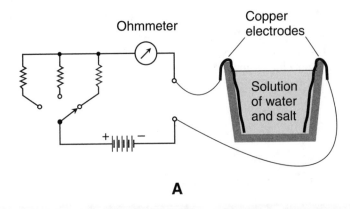

A

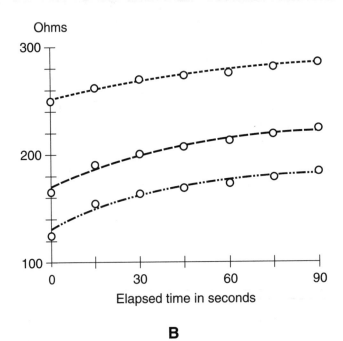

B

Figure DC7-2 *At A, the arrangement for measuring the resistance of water with salt (sodium chloride) fully dissolved. At B, the resistances I observed over a span of 90 seconds. Upper (short-dashed) curve: 0.5 teaspoon of salt. Middle (long-dashed) curve: 1.0 teaspoon of salt. Lower (dashed-and-dotted) curve: 1.5 teaspoons of salt.*

- After 60 seconds: 213 ohms

- After 75 seconds: 219 ohms

- After 90 seconds: 224 ohms

Once again, disconnect the positive probe wire from the meter. Add a third 0.5 teaspoon of salt to the solution, stir until it's dissolved, and let the solution settle. Do another series of timed measurements. Here are my results:

- At the beginning: 125 ohms

- After 15 seconds: 156 ohms

- After 30 seconds: 163 ohms

- After 45 seconds: 168 ohms

- After 60 seconds: 173 ohms

- After 75 seconds: 179 ohms

- After 90 seconds: 185 ohms

You should observe, as I did, a general decrease in the solution resistance as the salt concentration goes up, and an increase in the resistance over time with the ohmmeter connected. Figure DC7-2B is a multiple-curve graph of the data tabulated above.

Resistance of Soda Water

Once more, remove the positive probe jack from the meter outlet. Leave the ohmmeter range at the same setting. Empty the salt water from the cup. Rinse the cup and the electrodes. Fill the cup back up with the same amount of water as it contained before. Place the electrodes back in. Measure out precisely 0.5 teaspoon of baking soda (sodium bicarbonate) and pour it into the water. Stir until the soda is totally dissolved, and then allow the solution to calm down.

Refer to the clock once more. At the "top of the minute," plug the positive probe into the meter. Your system will now be interconnected as shown in Fig. DC7-3A. Measure the resistances at 15-second intervals. Here's what I observed:

- At the beginning: 473 ohms

- After 15 seconds: 738 ohms

- After 30 seconds: 847 ohms

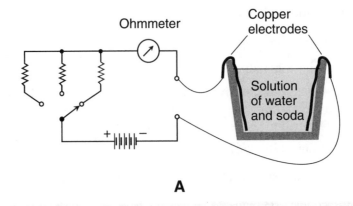

A

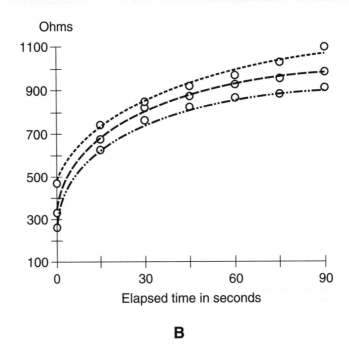

B

Figure DC7-3   *At A, the arrangement for measuring the resistance of water with soda (sodium bicarbonate) fully dissolved. At B, the resistances I observed over a span of 90 seconds. Upper (short-dashed) curve: 0.5 teaspoon of soda. Middle (long-dashed) curve: 1.0 teaspoon of soda. Lower (dashed-and-dotted) curve: 1.5 teaspoons of soda.*

- After 45 seconds: 918 ohms

- After 60 seconds: 977 ohms

- After 75 seconds: 1033 ohms

- After 90 seconds: 1096 ohms

Remove the positive probe from the meter receptacle, add another 0.5 teaspoon of soda, stir it until it's completely dissolved, and allow the solution to settle. Perform the timed measurements again. I got the following resistance values:

- At the beginning: 332 ohms

- After 15 seconds: 688 ohms

- After 30 seconds: 817 ohms

- After 45 seconds: 877 ohms

- After 60 seconds: 922 ohms

- After 75 seconds: 957 ohms

- After 90 seconds: 990 ohms

Disconnect the positive meter probe wire again. Add a third 0.5 teaspoon of soda to the solution, stir it in until it's completely dissolved, and let the solution settle. Reconnect the meter probe wire and conduct another set of timed measurements. Here are my results:

- At the beginning: 270 ohms

- After 15 seconds: 627 ohms

- After 30 seconds: 769 ohms

- After 45 seconds: 828 ohms

- After 60 seconds: 862 ohms

- After 75 seconds: 886 ohms

- After 90 seconds: 908 ohms

As before, you should see a decrease in the solution resistance as the soda concentration goes up, but an increase in the resistance as a function of time after the ohmmeter is connected. Figure DC7-3B is a multiple-curve graph of my results as tabulated above.

Why Does the Resistance Rise with Time?

In these experiments, *electrolysis* occurs because of the electric current driven through the solution by the ohmmeter. In electrolysis, the water (H_2O) molecules break apart into elemental hydrogen (H) and elemental oxygen (O), both of which are gases at room temperature. The gases accumulate as bubbles on the electrodes, some of which rise to the surface of the solution. However, both electrodes remain "coated" with some bubbles, which reduce the surface area of metal in contact with the liquid. That's why the apparent resistance of the solution goes up over time. If you stir the solution, you'll knock the bubbles off of the electrodes for a few moments, and the measured resistance will drop back down. If you let the solution come to rest again, the apparent resistance will rise once more as new gas bubbles accumulate on the electrodes.

DC8

Ohm's Law

In its basic form, *Ohm's law* states that the voltage across a component is directly proportional to the current it carries multiplied by its internal resistance. In this experiment, you'll demonstrate this law in two different ways. You'll need a size AA "flashlight" cell, a holder for the cell, a 330-ohm resistor, a 1000-ohm resistor, a 1500-ohm resistor, and your trusty current/voltage/resistance meter.

The Mathematics

Three simple equations define Ohm's law. If E represents the voltage in volts, I represents the current in amperes, and R represents the resistance in ohms, then

$$E = IR$$

If you know the voltage E across a component along with its internal resistance R, then you can calculate the current I through it as

$$I = E/R$$

If you know the voltage E across a component and the current I through it, then you can calculate its internal resistance R as

$$R = E/I$$

If any "inputs" are expressed in units other than volts, amperes, or ohms, then you must convert to those standard units before you begin calculations. Once you've done the arithmetic, you can convert the "output" to whatever unit you want (millivolts, microamperes, or kilohms, for example).

Check the Components!

Please let me recommend that you cultivate a lifelong habit: Test every electrical and electronic component *before* you put it to work in a circuit or system. If you check a component in isolation and find it bad, toss it aside immediately. If you let a defective component become part of a circuit or system, it will create havoc, and the problem may be harder to track down than a coyote in a canyon.

When I tested my AA cell with my digital voltmeter, I got a reading of 1.585 volts (V). When I tested the resistors with my digital ohmmeter, I observed 326 ohms, 981 ohms, and 1467 ohms. These values were well within the manufacturer's rated specifications. When you test your components, you should expect to get values that are a little (but not much) different from mine.

Current Determinations

Once the actual voltage and resistances have been determined, you're ready to predict how much current the cell will drive through each of the three resistors individually. Figure DC8-1 shows the arrangement for making measurements. Use the Ohm's law formula

$$I = E/R$$

where I is the predicted current, E is the known voltage, and R is the known resistance.

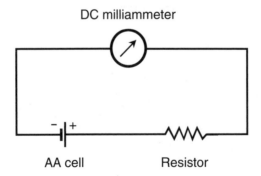

Figure DC8-1   *To measure the current that a known voltage drives through a known resistance, simply connect the voltage source, the resistor, and the meter in series.*

With the "330-ohm" resistor connected in series with the cell and the meter, I predicted that I would see a current value of

$$I = 1.585/326$$

$$= 0.00486 \text{ A}$$

$$= 4.86 \text{ mA}$$

When I made the measurement, I got 4.69 milliamperes (mA), which was 0.17 mA below the predicted current.

With the "1000-ohm" resistor connected in series with the cell and the meter, I predicted that I would observe

$$I = 1.585/981$$

$$= 0.00162 \text{ A}$$

$$= 1.62 \text{ mA}$$

I measured an actual current of 1.60 mA, which was 0.02 mA below the predicted value.

With the "1500-ohm" resistor connected in series with the cell and the meter, I predicted that I would see

$$I = 1.585/1467$$

$$= 0.00108 \text{ A}$$

$$= 1.08 \text{ mA}$$

I measured 1.08 mA, exactly the predicted value.

Experimental Error

In any laboratory environment, there's bound to be some *experimental error*. No instrument is perfect, and the real world is subject to uncertainty by nature. To determine the error in percent (%), do these calculations in order:

- Take the measured value

- Subtract the predicted value

- Divide by the predicted value

- Multiply by 100

- Round the answer off to the nearest whole number

Be sure you use the same units for the measured and predicted values. In the first case, my experimental error was

$$[(4.69 - 4.86)/4.86] \times 100$$
$$= (-0.17/4.86) \times 100$$
$$= -0.0350 \times 100$$
$$= -3.50\%$$

which rounds off to −4 percent. In the second case, my experimental error was

$$[(1.60 - 1.62)/1.62] \times 100$$
$$= (-0.02/1.62) \times 100$$
$$= -0.0123 \times 100$$
$$= -1.23\%$$

which rounds off to −1 percent. In the third case, my experimental error was

$$[(1.08 - 1.08)/1.08] \times 100$$
$$= (0.00/1.08) \times 100$$
$$= 0.00 \times 100$$
$$= 0.00\%$$

which rounds off to 0 percent. Of course, you should make your own calculations and derive your own results. Expect to get errors of a magnitude similar to those I experienced, either positive or negative.

If you use an analog meter, you must contend with *interpolation error* as well as experimental error. That's because you'll have to *interpolate* (make a good guess at) the position of the needle on the meter scale.

Voltage Determinations

The second Ohm's law experiment involves measuring the voltages across each of the three resistors when they're connected in series with the cell across the combination as shown in Fig. DC8-2. Here, you should use the Ohm's law formula

$$E = IR$$

where E is the predicted voltage, I is the measured current, and R is the measured resistance.

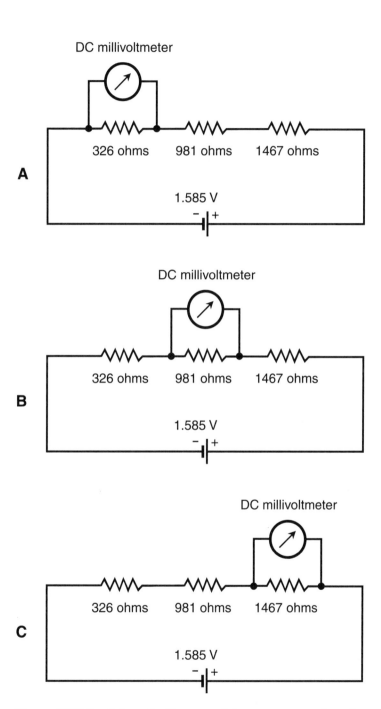

Figure DC8-2   *Schematic diagrams of the arrangements I used to measure the voltage produced by a current of 553 μA through a 326-ohm resistance (A), a 981-ohm resistance (B), and a 1467-ohm resistance (C).*

When I connected the three resistors in series with the DC milliammeter and the cell, I observed a current of 553 µA through the combination of resistors, equivalent to 0.000553 A. Across the "330-ohm" resistor (Fig. DC8-2A), I predicted that I would see a potential difference of

$$E = 0.000553 \times 326$$

$$= 0.180 \text{ V}$$

$$= 180 \text{ mV}$$

When I made the measurement, I got 186 mV, which was 6 mV more than the predicted potential difference. My experimental error was therefore

$$[(186 - 180)/180] \times 100$$

$$= (6/180) \times 100$$

$$= 0.0333 \times 100$$

$$= +3.33\%$$

$$\approx +3\%$$

The "wavy" equals sign indicates that the answer has been rounded off or approximated. You'll often see this symbol in papers, articles, and books involving experimental science. The plus sign indicates that my experimental error was positive in this case.

Across the "1000-ohm" resistor alone (Fig. DC8-2B), I predicted that I would see a potential difference of

$$E = 0.000553 \times 981$$

$$= 0.542 \text{ V}$$

$$= 542 \text{ mV}$$

I measured 560 mV, which was 18 mV more than the predicted potential difference. I calculated the experimental error as

$$[(560 - 542)/542] \times 100$$

$$= (18/542) \times 100$$

$$= 0.0332 \times 100$$

$$= +3.32\%$$

$$\approx +3\%$$

Across the "1500-ohm" resistor alone (Fig. DC8-2C), I predicted a potential difference of

$$E = 0.000553 \times 1467$$
$$= 0.811 \text{ V}$$
$$= 811 \text{ mV}$$

I measured 838 mV, which was 27 mV above the predicted potential difference. Calculating the experimental error, I got

$$[(838 - 811)/811] \times 100$$
$$= (27/811) \times 100$$
$$= 0.0333 \times 100$$
$$= +3.33\%$$
$$\approx +3\%$$

DC9

Resistors in Series

In this experiment, you'll see how the *ohmic values* (that's a fancy term for DC resistance) of resistors combine when you connect them in series. You'll need the same three resistors you used in Experiment DC8, along with an ohmmeter.

The Formula

When two or more components, each having a fixed DC resistance, are connected in series, their ohmic values add up. If R_1 and R_2 are the values of two resistors in ohms, then the total resistance R of the series combination, also in ohms, is

$$R = R_1 + R_2$$

This formula works for resistances in units other than ohms (such as kilohms or megohms), but only if you use the same unit all the way through.

If you have *n* resistors whose ohmic values are R_1, R_2, R_3, . . ., and R_n and you connect them all in series, then the total resistance R across the combination is

$$R = R_1 + R_2 + R_3 + \cdots + R_n$$

All of the resistances must be expressed in the same unit (ohms, kilohms, or megohms) if you want this formula to work. It doesn't matter how the resistors are arranged. You can connect them end-to-end in any sequence, and the total series resistance will always be the same.

Phase 1

When I did Experiment DC8, I measured the values of my resistors as 326 ohms, 981 ohms, and 1467 ohms. (Your resistors will have slightly different values, but they shouldn't be a lot different.) When I connected the 326-ohm resistor in series

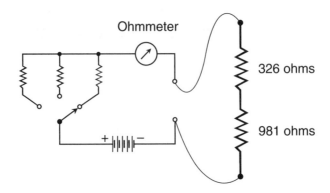

Figure DC9-1   *Series combination of 326 ohms and*
981 ohms.

with the 981-ohm resistor as shown in Fig. DC9-1, I expected to get a total resistance of

$$R = 326 + 981 = 1307 \text{ ohms}$$

I didn't want to use the breadboard for such a simple arrangement of components, so I simply twisted the resistors' leads together and then measured the total series resistance by holding the meter probe tips against the outer leads. I wore gloves to keep my body resistance from affecting the reading. With my digital ohmmeter set to the range 0 to 2000 ohms, I got a display reading of 1305 ohms, comfortably within the limits of experimental error. When you do the same thing with the two smallest resistors in your collection, you should get a similar small error, hopefully less than 1 percent. Be sure to use your measured values for the individual resistors, not the manufacturer's specified values of 330 and 980 ohms.

Phase 2

For the second part of the experiment, I connected my 326-ohm resistor in series with my 1467-ohm resistor as shown in Fig. DC9-2, again twisting the leads together and holding the meter probe tips against the outer leads. I expected to get a total resistance of

$$R = 326 + 1467 = 1793 \text{ ohms}$$

My meter, again set for the range 0 to 2000 ohms, displayed 1791 ohms. I was satisfied, once again, that the laws of basic electricity were still being obeyed. Your smallest and largest resistors should behave in the same orderly fashion when you connect them in series and measure the total resistance.

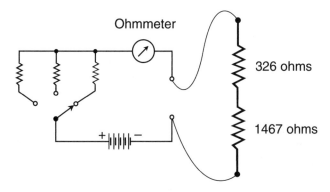

Figure DC9-2   *Series combination of 326 ohms and 1467 ohms.*

Phase 3

Phase 3 of the experiment involved the two largest resistors: the 981-ohm component and the 1467-ohm component. When I twisted the leads together and measured the total resistance as shown in Fig. DC9-3, I expected to obtain

$$R = 981 + 1467 = 2448 \text{ ohms}$$

My digital ohmmeter meter won't display this value when set for the range 0 to 2000 ohms, so I set it to the next higher range, which was 0 to 20 kilohms (K). The display showed 2.44, meaning 2.44 K. My predicted value, rounded off to the

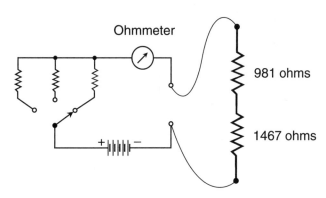

Figure DC9-3   *Series combination of 981 ohms and 1467 ohms.*

nearest hundredth of a kilohm, was 2.45 K. The combined experimental and measurement error was therefore

$$[(2.44 - 2.45)/2.45] \times 100$$

$$= (-0.01/2.45) \times 100$$

$$= -0.408\%$$

That's considerably less than plus-or-minus 1 percent (±1%), so this experiment worked out okay for me.

Phase 4

For the final part of this experiment, I connected all three of the resistors in series, in order of increasing resistance, by twisting the leads together. I placed the 326-ohm resistor on one end, the 981-ohm resistor in the middle, and the 1467-ohm resistor on the other end. Then I connected the ohmmeter as shown in Fig. DC9-4. According to my prediction, the total resistance should have been

$$R = 326 + 981 + 1467 = 2774 \text{ ohms}$$

With the ohmmeter set for the range 0 to 20 K, I got a reading of 2.77 on the digital display, meaning 2.77 K. When rounded off to the nearest hundredth of a kilohm, my prediction agreed precisely with the display.

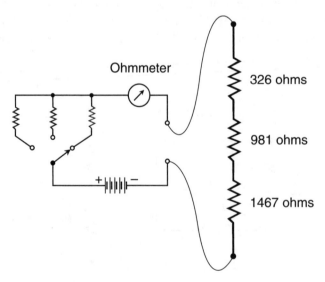

Figure DC9-4   *Series combination of 326 ohms, 981 ohms, and 1467 ohms.*

Now Try This!

Measure the total resistance of a series combination of these same three resistors, but with the largest resistance in the middle. Then do it again with the smallest resistance in the middle. Do you get the same results as you did when you connected the resistors end-to-end in ascending order of resistance? You should, within the bounds of measurement and experimental error.

DC10

Resistors in Parallel

In this experiment, you'll combine resistors in parallel and determine the net resistance, both theoretically and experimentally. You'll need the same resistors and meter that you used in Experiments DC8 and DC9.

What's a Siemens?

When two or more resistors are connected in parallel, their ohmic values combine in a rather complicated way. The formulas are straightforward enough, but they can be confusing. Students often have trouble understanding how the formulas work, and even veteran engineers don't always remember them.

Instead of thinking in terms of resistance, let's think in terms of *conductance*. Resistance is a measure of the extent to which a component or device *opposes* the flow of electric current. We express resistance in ohms. Conductance is a measure of the extent to which a component or device *allows* current to flow. We express conductance in units called siemens (symbolized S).

As the conductance of a particular component or device increases, its resistance decreases. As the resistance increases, the conductance decreases. The smallest possible conductance is 0 siemens (0 S), representing an open circuit. There's no limit to how large the conductance can get. A perfect short circuit would theoretically have a conductance of "infinity siemens," but in any practical situation, there's always a little resistance, so the conductance is extremely large but not infinite.

If you know the resistance of a component or circuit in ohms, you can get the conductance in siemens by taking the reciprocal. Conversely, if you know the conductance in siemens, you can calculate the resistance by taking the reciprocal. Resistance (as a mathematical variable) is denoted by an italicized, uppercase letter R. Conductance (as a variable) is denoted as an italicized, uppercase letter G. If we express R in ohms and G in siemens, then

$$G = 1/R$$

and

$$R = 1/G$$

Engineers and technicians often use units of conductance much smaller than the siemens. A resistance of 1 kilohm (1 K) is equivalent to 1 *millisiemens* (1 mS). If the resistance is 1 megohm (1 M), the conductance is 1 *microsiemens* (1 µS).

The Formula

In a parallel circuit, conductance values simply add, just as resistance values add in a series circuit. If G_1 and G_2 are the conductances of two components in siemens, then the total conductance G of the parallel combination, also in siemens, is

$$G = G_1 + G_2$$

This formula also works for conductances expressed in millisiemens or microsiemens, but only if you stay with the same unit throughout the calculation process. The addition rule also works for parallel combinations of three or more components. If you have n parallel-connected components whose conductances are G_1, G_2, G_3, . . ., and G_n, then the net conductance G across the combination is

$$G = G_1 + G_2 + G_3 + \cdots + G_n$$

Again, if you want this formula to work, then you must express all of the conductances in the same unit (siemens, millisiemens, or microsiemens).

Calculator Error

All digital calculators display a certain maximum number of digits. My calculator is a "dime-store" unit with a 10-digit display. It simply "chops off" all the digits after the 10th one. Mathematically, this process is called *truncation*. For improved accuracy, a more sophisticated calculator will leave the last displayed digit alone if the next digit would turn out as 0 to 4, and will increase the last displayed digit by 1 if the next one would turn out as 5 to 9. This process, called *rounding* or *rounding off*, provides better accuracy than truncation, especially when multiple calculations are done one after another.

To eliminate errors caused by the limited number of digits available on a calculator display, it's always a good idea to use a lot more digits during a calculation process than you'll need in the end. For example, if you want to obtain an answer that's accurate to three decimal places and you have a 10-digit calculator, you can

keep track of all the digits the calculator shows during each step of your arithmetic. Then, when everything is finished and you've found your final answer, you
can round it off (not truncate it) to three decimal places.

Phase 1

When I did the experiments described here, I used the same three resistors as I did
in the previous two experiments. Individually, they tested at 326 ohms, 981 ohms,
and 1467 ohms. When I connected the 326-ohm resistor in parallel with the
981-ohm resistor as shown in Fig. DC10-1 by twisting the leads together,
I expected to get a total conductance of

$$G = (1/326 + 1/981) \text{ S}$$

Using my calculator, I worked out this value as

$$G = 0.003067484 + 0.001019367$$

$$= 0.004086851 \text{ S}$$

This conductance is equivalent to a resistance of

$$R = 1/G$$

$$= 1/0.004086851$$

$$= 245 \text{ ohms}$$

rounded off to the nearest ohm. When I measured the total series resistance with
my meter set to the range 0 to 2000 ohms, the display showed exactly this value.

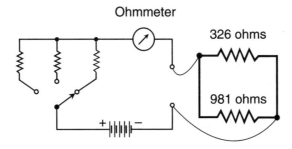

Figure DC10-1   *Parallel combination of 326-ohm
and 981-ohm resistances.*

Phase 2

For the second part of the experiment, I connected the 326-ohm resistor in parallel with the 1467-ohm resistor as shown in Fig. DC10-2. Converting the resistances to conductances and then adding them together, I got a predicted value of

$$G = 1/326 + 1/1467$$
$$= 0.003067484 + 0.000681663$$
$$= 0.003749147 \text{ S}$$

This conductance is the same as a resistance of

$$R = 1/G$$
$$= 1/0.003749147$$
$$= 267 \text{ ohms}$$

rounded off to the nearest ohm. My ohmmeter displayed exactly that!

Phase 3

Next, I connected the 981-ohm resistor in parallel with the 1467-ohm resistor as shown in Fig. DC10-3. When I converted these ohmic values to conductances and then added them, I came up with

$$G = 1/981 + 1/1467$$
$$= 0.001019367 + 0.000681663$$
$$= 0.001701030 \text{ S}$$

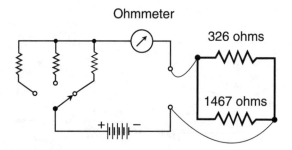

Figure DC10-2   *Parallel combination of 326-ohm and 1467-ohm resistances.*

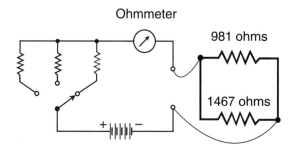

Figure DC10-3   *Parallel combination of 981-ohm and 1467-ohm resistances.*

This conductance is equivalent to a resistance of

$$R = 1/G$$

$$= 1/0.001701030$$

$$= 588 \text{ ohms}$$

rounded off to the nearest ohm. When I made the measurement, my ohmmeter rewarded me by displaying this exact number.

Phase 4

Finally, I connected all three resistors in parallel as shown in Fig. DC10-4. When I twisted the wire leads of my three resistors together to make a messy but workable parallel circuit, I expected a total conductance of

$$G = 1/326 + 1/981 + 1/1467$$

$$= 0.003067484 + 0.001019367 + 0.000681663$$

$$= 0.004768514 \text{ S}$$

which came out to a predicted resistance of

$$R = 1/G$$

$$= 1/0.004768514$$

$$= 210 \text{ ohms}$$

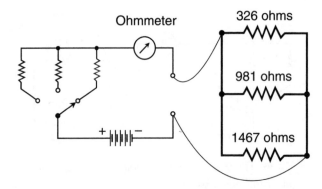

Figure DC10-4   *Parallel combination of 326-ohm, 981-ohm, and 1467-ohm resistances.*

rounded off to the nearest ohm. With the ohmmeter still set to the range 0 to 2000 ohms, I got a reading that was right on target. I was amazed: I had done four experiments in a row, and had suffered from discernible error in none of them.

Ponder This!

Compare the errors in the results of this experiment with the errors in the measurements you made during Experiment DC9. Do you suspect that parallel circuits might be inherently more well-behaved than series circuits?

DC11

Resistors in Series-Parallel

In this experiment, you'll observe how resistances combine when they're connected in *series-parallel networks*. Reuse the resistors from Experiments DC8, DC9, and DC10. Get ready for some repetitive lab work!

Phase 1

Connect the two larger resistors in series, and then connect the smallest one across that combination. After I wired my 981-ohm and 1467-ohm resistors in series by twisting the leads together, I calculated the resistance of the combination and called it R_1, getting

$$R_1 = 981 + 1467$$
$$= 2448 \text{ ohms}$$

Next, I called my 326-ohm resistance R_2, wired it in parallel with the series combination R_1 as shown in Fig. DC11-1 (again by twisting the component leads together), and figured that the complete series-parallel network should have a net conductance of

$$G = 1/R_1 + 1/R_2$$
$$= 1/2448 + 1/326$$
$$= 0.000408496 + 0.003067484$$
$$= 0.003475980 \text{ S}$$

My prediction for the total network resistance was therefore

$$R = 1/G$$
$$= 1/0.003475980$$
$$= 288 \text{ ohms}$$

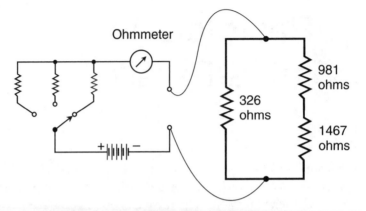

Figure DC11-1   *Resistive network with 326 ohms across a series combination of 981 ohms and 1467 ohms.*

rounded off to the nearest ohm. When I measured R with my ohmmeter switched to read values in the range 0 to 2000 ohms, the display showed 287 ohms.

Phase 2

Connect the smallest and largest resistors in series, and calculate the theoretical net resistance R_1. I predicted

$$R_1 = 326 + 1467$$
$$= 1793 \text{ ohms}$$

Calling the 981-ohm resistor R_2 and connecting it across the series combination R_1 as shown in Fig. DC11-2, I predicted a net conductance of

$$G = 1/R_1 + 1/R_2$$
$$= 1/1793 + 1/981$$
$$= 0.000557724 + 0.001019367$$
$$= 0.001577091 \text{ S}$$

My calculated total network resistance was

$$R = 1/G$$
$$= 1/0.001577091$$
$$= 634 \text{ ohms}$$

I measured an actual value of $R = 633$ ohms.

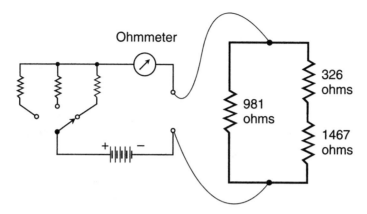

Figure DC11-2 *Resistive network with 981 ohms across a series combination of 326 ohms and 1467 ohms.*

Phase 3

Connect the two smaller resistors in series and calculate the theoretical net resistance R_1. I predicted

$$R_1 = 326 + 981$$
$$= 1307 \text{ ohms}$$

Calling the 1467-ohm resistor R_2 and placing it across R_1 as shown in Fig. DC11-3, I predicted a net conductance of

$$G = 1/R_1 + 1/R_2$$
$$= 1/1307 + 1/1467$$
$$= 0.000765110 + 0.000681663$$
$$= 0.001446773 \text{ S}$$

My calculated total network resistance was

$$R = 1/G$$
$$= 1/0.001446773$$
$$= 691 \text{ ohms}$$

I measured an actual value of $R = 691$ ohms.

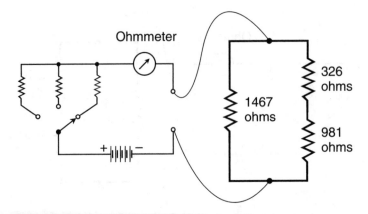

Figure DC11-3   *Resistive network with 1467 ohms across a series combination of 326 ohms and 981 ohms.*

Phase 4

Now let's change the basic geometry of the series-parallel network and do three more experiments. Connect the two larger resistors in parallel and calculate the theoretical net resistance R_1. I predicted a conductance of

$$G = 1/981 + 1/1467$$
$$= 0.001019367 + 0.000681663$$
$$= 0.001701030 \text{ S}$$

which is a net resistance of

$$R_1 = 1/G$$
$$= 1/0.001701030$$
$$= 588 \text{ ohms}$$

Calling the 326-ohm resistor R_2 and wiring it in series with R_1 as shown in Fig. DC11-4, I predicted a total network resistance of

$$R = R_1 + R_2$$
$$= 588 + 326$$
$$= 914 \text{ ohms}$$

I measured an actual value of $R = 913$ ohms.

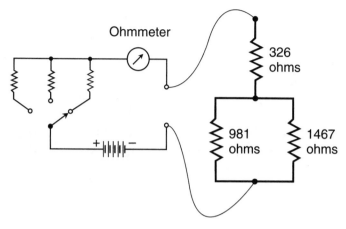

Figure DCII-4 *Resistive network with 326 ohms in series with a parallel combination of 981 ohms and 1467 ohms.*

Phase 5

Connect the smallest resistor in parallel with the largest one and calculate the theoretical net resistance R_1. I predicted a conductance of

$$G = 1/326 + 1/1467$$
$$= 0.003067484 + 0.000681663$$
$$= 0.003749147 \text{ S}$$

which is a net resistance of

$$R_1 = 1/G$$
$$= 1/0.003749147$$
$$= 267 \text{ ohms}$$

Calling the 981-ohm resistor R_2 and wiring it in series with R_1 as shown in Fig. DC11-5, I predicted a total network resistance of

$$R = R_1 + R_2$$
$$= 267 + 981$$
$$= 1248 \text{ ohms}$$

I measured an actual value of $R = 1246$ ohms.

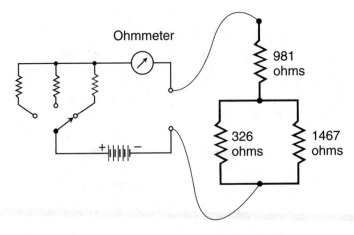

Figure DC11-5   *Resistive network with 981 ohms in series*
with a parallel combination of 326 ohms and 1467 ohms.

Phase 6

Connect the two smallest resistors in parallel and calculate the theoretical net resistance R_1. I predicted a conductance of

$$G = 1/326 + 1/981$$
$$= 0.003067484 + 0.001019367$$
$$= 0.004086851 \text{ S}$$

which is a net resistance of

$$R_1 = 1/G$$
$$= 1/0.004086851$$
$$= 245 \text{ ohms}$$

Calling the 1467-ohm resistor R_2 and wiring it in series with R_1 as shown in Fig. DC11-6, I predicted a total network resistance of

$$R = R_1 + R_2$$
$$= 245 + 1467$$
$$= 1712 \text{ ohms}$$

I measured an actual value of $R = 1708$ ohms.

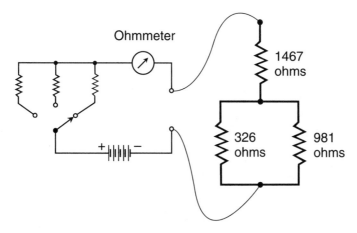

Figure DC11-6 *Resistive network with 1467 ohms in series with a parallel combination of 326 ohms and 981 ohms.*

Ponder This!

You'll notice that my predictions were perfect in some instances, one or two digits off in others, and four digits off in the final case. When you do these experiments, you might also get varying amounts of error. Can you explain why some parts of this experiment work out better than others, when all the components are identical in every case?

DC12

Kirchhoff's Current Law

In this experiment, you'll construct a network that demonstrates one of the most important principles in DC electricity. You'll need five fresh resistors: two rated at 330 ohms, one rated at 1000 ohms, and two rated at 1500 ohms. You'll also need a lantern battery, or a set of four AA cells connected in series.

Use the Breadboard!

So far, you haven't had to use the breadboard described in "Your Direct-Current Lab" at the beginning of Part 1. Things will go more smoothly now if you take advantage of it. Mount the five resistors by wire-wrapping the leads around the terminal nails as shown in Fig. DC12-1. Remember to test each resistor with your ohmmeter to verify their actual values before you install them. Use a 5-inch (5-in) long bare copper wire to interconnect the three terminals I-1, J-1, and K-1. Do the same with I-3, J-3, and K-3, and also with I-5, J-5, and K-5.

Abbreviations

Engineers symbolize current *as a variable* (in equations, for example) by writing an uppercase italic letter *I*. Amperes *as units* are abbreviated as an uppercase nonitalic letter A. Milliamperes are abbreviated as mA, and microamperes are abbreviated as μA.

Voltage *as a variable* can be symbolized by an uppercase italic letter *E* or an uppercase italic letter *V*. You should get used to seeing it both ways. Volts *as units* are abbreviated as an uppercase nonitalic V. Millivolts are abbreviated as mV, microvolts as μV, kilovolts as kV, and megavolts as MV.

Resistance *as a variable* is symbolized with the uppercase italic letter *R*. Ohms *as units* can be written out in full, although some texts use the uppercase nonitalic Greek letter omega (Ω). Kilohms are abbreviated as K ohms or K. Megohms are abbreviated as M ohms or M.

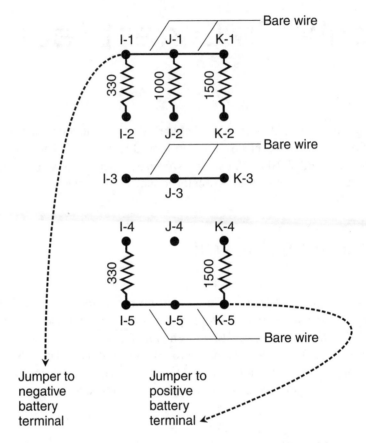

Figure DC12-1   *Arrangement of resistors on breadboard for demonstration of Kirchhoff's current law. All resistance values are in ohms. Solid dots indicate breadboard terminals. Solid lines show interconnections with bare copper wire. Dashed lines indicate jumpers.*

What Does the Law Say?

Gustav Robert Kirchhoff (1824–1887) did research and formulated theories in a time when little was known about electrical current. He used common sense to deduce fundamental properties of DC circuits.

Kirchhoff reasoned that the current going into any *branch point* in a circuit is always the same as the current going out of that point. Figure DC12-2 is a generic example of this principle, known as *Kirchhoff's first* law. We can also call it *Kirchhoff's current law* or the principle of *conservation of current*.

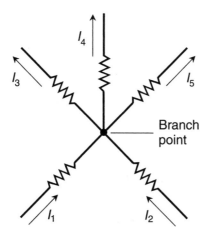

Figure DC12-2 *According to Kirchhoff's current law, the sum of the currents flowing into any branch point is equal to the sum of the currents flowing out of that branch point. In this example,*
$I_1 + I_2 = I_3 + I_4 + I_5.$

Mathematically, the sum of the currents entering a branch point is the same as the sum of the currents leaving it. In the example of Fig. DC12-2, two branches enter the point and three branches leave it, so

$$I_1 + I_2 = I_3 + I_4 + I_5$$

Kirchhoff's current law holds true no matter how many branches come in or go out of a particular point.

Do the Tests

Connect the battery to the resistive network and measure the currents in each branch. Every test point should be metered individually, while all the other test points are shorted with jumpers. Figure DC 12-3 illustrates the actual values of the resistors in my network (yours will be slightly different, of course), along with the value I got when I measured I_1, the current through the smaller of the two input resistors.

As you measure each of the four other current values in turn, be sure that the meter polarity always agrees with the battery polarity. The black meter probe

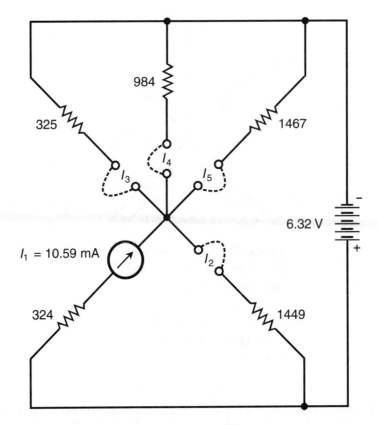

Figure DC12-3   *Network for verifying Kirchhoff's current law.
All resistance values are in ohms. The battery voltage, the current
I_1, and the resistances are the values I measured. Dashed lines
show interconnections with jumpers.*

should go to the more negative point, and the red meter probe should go to the
more positive point. That way, you'll avoid getting negative current readings that
might throw off your calculations. When I measured I_1 through I_5, I got these
results, accurate to the nearest hundredth of a milliampere:

$$I_1 = 10.59 \text{ mA}$$
$$I_2 = 2.40 \text{ mA}$$
$$I_3 = 8.35 \text{ mA}$$
$$I_4 = 2.79 \text{ mA}$$
$$I_5 = 1.88 \text{ mA}$$

It's "mission critical" that all test points *not* undergoing current measurement be
shorted out with jumpers. Otherwise, your network will be incomplete and your

current measurements will come out wrong. After you've finished making measurements, remove all jumpers to conserve battery energy.

Calculate the Sums

"Plug in the numbers" to the Kirchhoff formulas and see how close the sum of the input currents comes to the sum of the output currents. Here are my results for the sum of the currents entering the branch point:

$$I_1 + I_2 = 10.59 + 2.40$$
$$= 12.99 \text{ mA}$$

When I added the currents leaving the branch point, I got

$$I_3 + I_4 + I_5 = 8.35 + 2.79 + 1.88$$
$$= 13.02 \text{ mA}$$

Now Try This!

Use Ohm's law, along with the rules for resistance combinations in series and parallel, to "retroactively predict" the current measurements you just made. How close do the theoretical values come to your experimental outcomes?

DC13

Kirchhoff's Voltage Law

In this experiment, you'll construct a network that demonstrates another important DC circuit rule. You'll need four fresh resistors: one rated at 220 ohms, one rated at 330 ohms, one rated at 470 ohms, and one rated at 680 ohms. You'll also need a lantern battery, or a set of four AA cells connected in series.

What Does the Law Say?

According to *Kirchhoff's second law*, the sum of the voltages across the individual components in a series DC circuit, taking polarity into account, is always equal to zero. We can also call this rule *Kirchhoff's voltage law* or the principle of *conservation of voltage*.

Consider the generic series DC circuit shown in Fig. DC13-1. According to Kirchhoff, the battery voltage, E, should be the same as the sum of the potential differences across the resistors, although the polarity will be reversed. Mathematically, we can state this fact as

$$E + E_1 + E_2 + E_3 + E_4 = 0$$

If we measure the voltages across the individual resistors and the battery, one at a time, with a DC voltmeter and *disregard the polarity*, we should find that

$$E = E_1 + E_2 + E_3 + E_4$$

Do the Tests

Check each of the four resistors with your ohmmeter to verify their actual values. Mount the resistors in the upper right-hand corner of your breadboard by wire-wrapping the leads around nails as shown in Fig. DC13-2. Connect the battery to the network as shown, and measure the voltage across each resistor. Figure DC13-3 illustrates the actual values of the resistors in my network (yours will be slightly

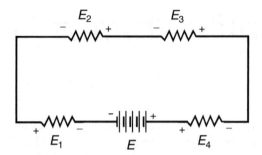

Figure DC13-1 *According to Kirchhoff's voltage law, the sum of the voltages across the resistances in a series DC circuit is equal and opposite to the battery voltage. If we disregard polarity, then in this example, we'll observe that $E = E_1 + E_2 + E_3 + E_4$.*

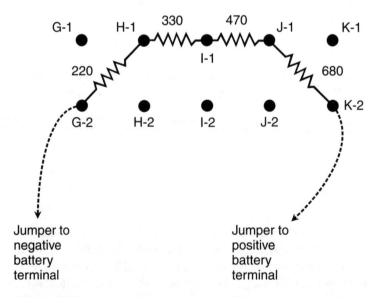

Figure DC13-2 *Suggested arrangement of resistors on breadboard for demonstration of Kirchhoff's voltage law. All resistance values are in ohms. Solid dots indicate terminals. Dashed lines indicate jumpers.*

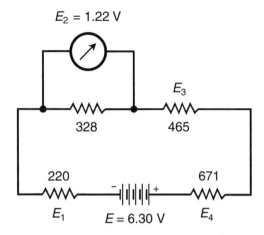

$E_2 = 1.22$ V

E_3

328   465

220   671

E_1 $E = 6.30$ V E_4

Figure DC13-3   *Network for verifying Kirchhoff's voltage law. All resistance values are in ohms. The battery voltage E, the voltage across the second resistor, and the resistances are the values I measured.*

different), along with the voltage I got for E_2. I measured $E = 6.30$ volts (V) across the battery when it was under load.

As you measure each voltage E_1 through E_4, the black meter probe should go to the more negative voltage point, and the red probe should go to the more positive point. That way, you'll avoid getting negative readings that might throw off your calculations. When I measured the voltages across the individual resistors, I got

$$E_1 = 0.82 \text{ V}$$
$$E_2 = 1.22 \text{ V}$$
$$E_3 = 1.75 \text{ V}$$
$$E_4 = 2.52 \text{ V}$$

After you've finished making your measurements, remove one of the jumpers to take the load off the battery.

Calculate the Sums

Once you've double-checked and written down your voltage measurements, input the numbers to the modified Kirchhoff formula

$$E = E_1 + E_2 + E_3 + E_4$$

and see how closely it works out. For the left-hand side of this equation, I measured

$$E = 6.30 \text{ V}$$

and for the right-hand side, I added my numbers to get

$$E_1 + E_2 + E_3 + E_4 = 0.82 + 1.22 + 1.75 + 2.52$$
$$= 6.31 \text{ V}$$

Now Try This!

Set your meter to measure DC milliamperes, and replace one of the jumper wires in Fig. DC13-2 with it. Measure the current drawn from the battery, and then use Ohm's law to "retroactively predict" the voltage across each individual resistor. How close do these theoretical values come to the results you got when you measured the voltages?

DC14

A Resistive
Voltage Divider

For this experiment, you'll use the components from Experiment DC13 to obtain several different voltages from a single battery. Keep the resistors on the breadboard in the same arrangement as you had them in Experiment DC13.

How Does It Work?

When two or more resistors are connected in series with a DC power source, those resistors produce specific voltage ratios. These ratios can be tailored using a *voltage divider* that "fixes" the intermediate voltages. This type of circuit works best when the resistance values are fairly small.

Figure DC14-1 illustrates the principle of a *resistive voltage divider*. The individual resistances are R_1, R_2, R_3, . . . , and R_n. The total resistance R is the sum

$$R = R_1 + R_2 + R_3 + \cdots + R_n$$

If we call the supply voltage E, then Ohm's law tells us that the current I at any point in the circuit must be

$$I = E/R$$

as long as we express I in amperes (A), E in volts (V), and R in ohms. At the points P_1, P_2, P_3, . . . , and P_n, the voltages relative to the negative battery terminal are E_1, E_2, E_3, . . . , and E_n, respectively. The last (and highest) voltage, E_n, is the same as the battery voltage, E. The voltages at the various points increase according to the sum total of the resistances up to each point, in proportion to the total resistance, multiplied by the supply voltage. In theory, we should find that the following equations hold true:

$$E_1 = ER_1/R$$
$$E_2 = E(R_1 + R_2)/R$$
$$E_3 = E(R_1 + R_2 + R_3)/R$$
$$\downarrow$$
$$E_n = E(R_1 + R_2 + R_3 + \cdots + R_n)/R = ER/R = E$$

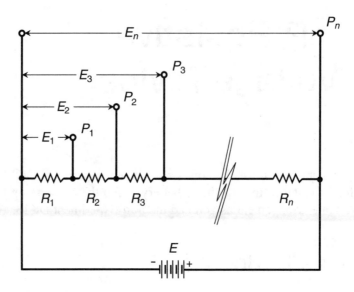

Figure DC14-1 *A voltage divider takes advantage of the potential differences across individual resistors connected in series with a DC power source.*

Measure the Current

During this experiment, I measured $E = 6.30$ V across the battery as it worked under load as shown in Fig. DC14-2A. This diagram shows the *rated* values of the resistors. Your actual values will, of course, be a little different than the rated values. In my case, they were

$$R_1 = 220 \text{ ohms}$$
$$R_2 = 328 \text{ ohms}$$
$$R_3 = 465 \text{ ohms}$$
$$R_4 = 671 \text{ ohms}$$

Set your meter for milliamperes (mA). Connect the battery to the resistive network through the meter as shown in Fig. DC14-2B, and measure the current. In theory, I expected the milliammeter to indicate a value equal to the battery voltage divided by the sum of the actual resistances, or

$$I = E/R$$
$$= 6.30/(220 + 328 + 465 + 672)$$
$$= 6.30/1684$$
$$= 0.00374 \text{ A}$$
$$= 3.74 \text{ mA}$$

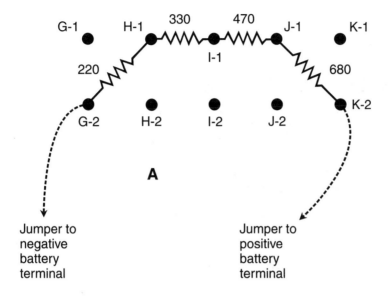

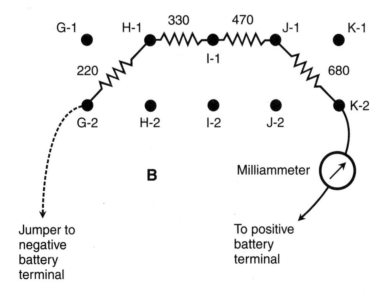

Figure DC14-2   *At A, suggested arrangement for measurements of the voltages across individual resistors. All resistance values are in ohms. Solid dots indicate terminals. Dashed lines indicate jumpers. At B, suggested arrangement for measurement of current through the network.*

When I measured the current, I got 3.73 mA, a value comfortably within the range of acceptable error.

Measure the Voltages

Measure the intermediate voltages E_1 through E_4 with your meter set for a moderate DC voltage range. The black meter probe should go directly to the negative battery terminal and stay there. The red meter probe should go to each positive voltage point in turn. First measure the voltage E_1 that appears across R_1 only. Then measure, in order, the following:

- The potential difference E_2 across $R_1 + R_2$
- The potential difference E_3 across $R_1 + R_2 + R_3$
- The potential difference E_4 across $R_1 + R_2 + R_3 + R_4$

Figure DC14-3 illustrates the arrangement for measuring E_2. My meter displayed these results:

$$E_1 = 0.82 \text{ V}$$
$$E_2 = 2.04 \text{ V}$$
$$E_3 = 3.79 \text{ V}$$
$$E_4 = 6.30 \text{ V}$$

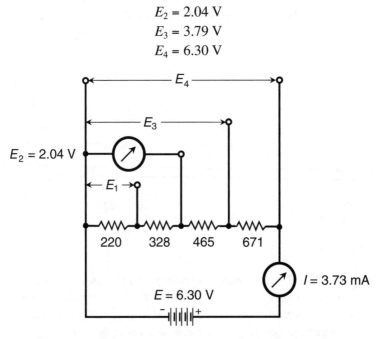

Figure DC14-3 *Network for testing the operation of a resistive voltage divider. All resistance values are in ohms. The battery voltage E, the voltage across the first and second resistors, and the resistances are those I measured.*

After you've finished making the voltage measurements, remove one of the jumpers to conserve battery energy.

Draw a Graph

Connect the meter across the combination $R_1 + R_2$. Run a couple of jumper wires to a *load resistor* located elsewhere on the breadboard, as shown in Fig. DC14-4.

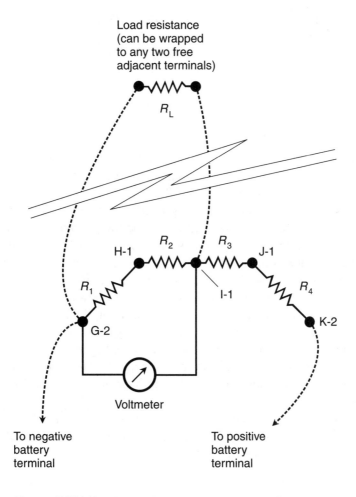

Figure DC14-4   *Circuit for testing a resistive voltage divider under load. Dashed lines indicate jumpers. This diagram shows the arrangement for measuring variations in E_2 as the load resistance R_L is alternately connected and disconnected from the series combination of R_1 and R_2.*

Table DC14-1   Voltages I measured across various loads in a voltage divider constructed according to Fig. DC14-4. My network resistor values were R_1 = 220 ohms, R_2 = 328 ohms, R_3 = 465 ohms, and R_4 = 671 ohms.

Load Resistance (ohms)	Output Voltage (volts)
No load	2.04
3.25 K	1.83
1470	1.63
983	1.49
671	1.32
466	1.14
326	0.96
220	0.76

This causes the voltage source E_2 to force current through the load resistor, which we'll call R_L. Try every resistor in your repertoire in the place of R_L. If you obtained all the resistors in the parts list (Table DC1-1), you'll have seven tests to do, using resistors rated at values ranging from 220 ohms to 3.3 kilohms (K).

Alternately connect and disconnect one of the jumper wires between the voltage divider and R_L, so that you can observe the effect of the load on E_2. As you can see, the behavior of a resistive voltage divider is affected by an external load. As R_L decreases, so does E_2. The effect becomes dramatic when R_L becomes small, representing a "heavy load." Table DC14-1 shows the results I got. Plot your results as points on a coordinate grid with R_L on the horizontal axis and E_2 on the vertical axis, and then "connect the dots" to get a *characteristic curve* showing the voltage as a function of the load resistance. Figure DC14-5 is the graph I made. I used a reverse logarithmic scale to portray R_L. This graph scale allows for a graph that provides a clear picture of what happens as the *conductance* of the load *increases*.

What do you think will happen to the voltage across the load if you use two, three, or four 220-ohm resistors in parallel, getting R_L values of about 110 ohms, 73 ohms, and 55 ohms, respectively?

So What?

The results of this experiment suggest that when engineers build voltage dividers, they had better know what sort of external load the circuit will have to deal with.

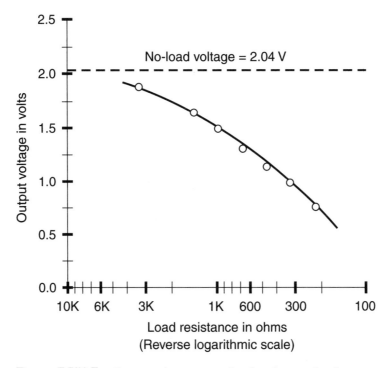

Figure DC14-5 *Output voltage versus load resistance for the resistive voltage divider. These are my results. The dashed line shows the open-circuit (no-load) voltage. Open circles show measured voltages under various loads. The solid curve approximates the circuit's characteristic function.*

If the load resistance fluctuates greatly, especially if it sometimes gets low, a resistive voltage divider won't work very well. There's another way to "tailor" the output of a DC power supply or battery that provides a more stable voltage under variable loads. You'll build one in the next experiment.

A Diode-Based Voltage Reducer

Rectifier diodes can be employed to reduce the output voltage of a DC battery or power supply. For this experiment, you'll need two *rectifier diodes*. The ones I obtained were rated at 1 ampere (A) and 600 *peak inverse volts* (PIV), available at Radio Shack stores as part number 276-1104. You'll also need at least one of each of the resistors listed in Table DC1-1, along with some jumpers.

How Does It Work?

Figure DC15-1 shows the schematic symbol for a diode, which is manufactured by joining a piece of *P-type* semiconductor material to a piece of *N-type* material. The N-type semiconductor, represented by the short, straight line, forms the diode's *cathode*. The P-type semiconductor, represented by the arrow, composes the *anode*.

Under most conditions, electrons can travel easily from the cathode to the anode (in the direction opposite the arrow), but not from the anode to the cathode (in the direction of the arrow). If you connect a battery and a resistor in series with the diode, current will flow if the negative terminal of the battery is connected to the cathode and the positive terminal is connected to the anode, as shown in Fig. DC15-2A. This condition is called *forward bias*. No current will flow if the battery is reversed, as shown in Fig. DC15-2B (except when the voltage is very high). This condition is called *reverse bias*. The resistor prevents destruction of the diode by excessive current.

A certain minimum voltage is necessary for current to flow through a forward-biased semiconductor diode. This "threshold" voltage is called the *forward breaker voltage*. In most diodes, it's a fraction of a volt, but it varies somewhat depending on how much current the diode is forced to carry.

If the forward-bias voltage across the diode's *P-N junction* is not at least as great as the forward breaker voltage, the diode will not conduct. When a diode is

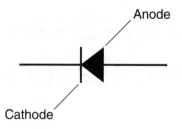

Figure DC15-1 *Schematic symbol for a semi-conductor diode. The short line represents the cathode. The arrow represents the anode.*

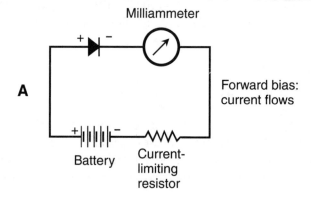

Figure DC15-2 *Series connection of a battery, a resistor, a current meter, and a diode. At A, forward bias causes current to flow if the voltage is at least equal to the forward breakover. At B, reverse bias results in no current through the diode, unless the voltage is very high.*

forward-biased and connected in series with a battery, the battery voltage is reduced to an extent approximately equal to the forward breakover voltage.

Construct and Test It

You can set up a voltage reducer with two diodes as shown in Fig. DC15-3, so that current flows through the load resistor R_L. Set your meter to indicate DC voltage in a moderate range such as 0 to 20 volts (V). Connect the meter across the load resistance, paying attention to the polarity so you get positive voltage readings. As you did in Experiment DC14, try every resistor you have for R_L. Measure the voltage across R_L in each case. You'll have seven tests to do, with resistances ranging from 220 ohms to 3.3 kilohms (K).

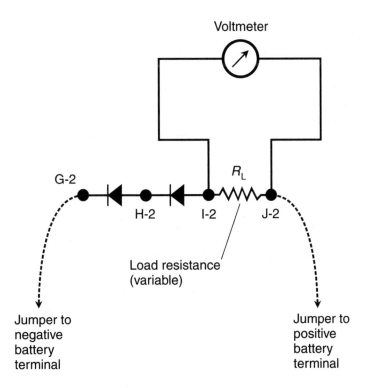

Figure DC15-3   *Suggested arrangement for measurements of the voltages across the load resistance in a two-diode voltage reducer. Solid dots show breadboard terminals. Dashed lines indicate jumpers. Pay attention to the diode polarity! The cathodes should go toward the negative battery terminal.*

Table DC15-1   Outputs I obtained with various loads connected to a diode-based voltage reducer. The circuit consisted of two diodes rated at 1 A and 600 PIV, forward-biased and placed in series with a 6.30-V battery.

Load Resistance (ohms)	Output Voltage (volts)
No load	5.70
3.25 K	5.08
1470	4.99
983	4.96
671	4.91
466	4.88
326	4.84
220	4.79

The load resistance affects the behavior of a diode-based voltage reducer, but in a different way than it affects the behavior of a resistive voltage divider. When you do these tests, you'll see that as the load resistance R_L decreases, the potential difference across it goes down—but not much. The voltage across the load tends to drop *more and more slowly* as R_L decreases. Contrast this with the resistive divider, in which the voltage drops off *more and more rapidly* as the load resistance goes down. Table DC15-1 shows the results I obtained when I measured the voltages across various loads.

Plot the Characteristic Curve

Plot your results as points on a coordinate grid with the load resistance on the horizontal axis and the voltage across the load resistor on the vertical axis, and then approximate the curve as you did in Experiment DC14. When I did this, I got the graph shown in Fig. DC15-4. As before, I used a reverse logarithmic scale to portray R_L. Compare this graph with Fig. DC14-5.

Now Try This!

Repeat this experiment with only one diode instead of two. Then, if you're willing to spend some more money and make another trip to Radio Shack, get another package of diodes and try the experiment with three or four of them in series. You might also obtain some more resistors, covering a range of values of, say, 100 ohms to 100 K, and test the circuit using those resistors in the place of R_L.

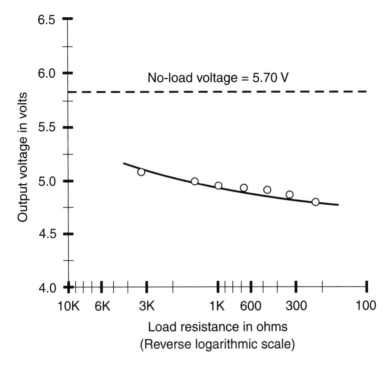

Figure DC15-4   *Output voltage versus load resistance for the diode-based voltage reducer. These are my results. The dashed line shows the open-circuit (no-load) voltage. Open circles show measured voltages under various loads. The solid curve approximates the circuit's characteristic function.*

Be careful here! Don't use a resistor of less than about 75 ohms as the load. In this arrangement, a ½-watt resistor of less than 75 ohms will let too much current flow, risking destruction of the resistor (and possibly the diodes as well).

Power as Volt-Amperes

For this experiment, you'll need two small direct-current (DC) incandescent lamps. The bulbs I obtained were rated at 6.3 and 7.5 volts (V) DC, available at Radio Shack stores as part numbers 272-1130 and 272-1133, respectively. You'll also need a couple of jumpers, four fresh (not old!) AA cells, and your breadboard with the battery holders and lamp holders securely mounted.

What Are DC Volt-Amperes?

We can calculate the power consumed by any component in a simple DC circuit such as any of those shown in this experiment. All we need to do is measure the voltage E across the component, measure the current I through it, and then multiply the voltage by the current to get *volt-ampere* (VA) *power* (P_{VA}). The formula is

$$P_{VA} = EI$$

where E is in volts, I is in amperes, and P_{VA} is an indirect expression of the power in watts consumed by the component or device.

Direct measurement of DC wattage or *true power* is more complicated than the measurement of VA power, and is beyond anything we can do with the equipment described in this book. To directly measure the true power consumed by the lamp in this circuit, we would have to determine how much total energy the lamp radiates over a period of time. We would also have to assume that all of the power supplied to the lamp actually comes out as light and heat, and not in any other form such as mechanical motion, X rays, or radio waves.

In simple DC circuits such as those described in this experiment, VA power is an excellent approximation of true power. But in more complex circuits such as those involving motors, generators, or alternating current (AC), the VA power can, and often does, differ from the true power.

Lamp Number One

Connect the 6.3-V lamp directly to a series combination of four AA cells using jumpers. The lamp should burn at full brilliance. Measure the potential difference across the lamp as shown in Fig. DC16-1A. (You shouldn't need breadboard-layout diagrams to interconnect the components in simple situations like this, so I haven't drawn a pictorial here.) Repeat the experiment with two AA cells in series,

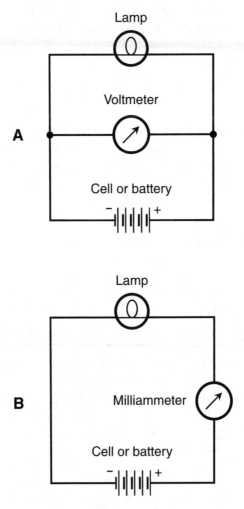

Figure DC16-1 *At A, the arrangement for measuring the voltage across a lamp. At B, the arrangement for measuring the current through a lamp.*

Table DC16-1   Voltages, currents, and VA power values I obtained by connecting one, two, and four AA cells in series to illuminate a 6.3-V incandescent lamp.

Number of AA Cells in Series	Volts across Lamp	Amperes through Lamp	Power as Volt-Amperes
1	1.25	0.093	0.116
2	2.52	0.140	0.353
4	5.85	0.250	1.46

and finally with only one AA cell. With two cells, of course, the lamp will burn less brightly, and with only one cell it will hardly glow at all. Compile the voltages in tabular form. When I did these experiments, I got the numbers in the second column of Table DC16-1.

Set your meter to measure DC amperes or milliamperes. Connect the lamp to the combinations of cells through your meter as shown in Fig. DC16-1B. Do the tests with four, two, and one AA cell. In each case, the lamp brilliance should be the same as it was during the corresponding voltage test: bright with four cells, dim with two cells, and hardly aglow with one cell. Compile the currents and enter them in tabular form. My results are shown in the third column of Table DC16-1.

Once you've obtained the lamp voltages and currents using the three different batteries, multiply volts by amperes to get VA power values. If you made measurements in millivolts or milliamperes, you must convert them to volts and amperes before using the formula. My results appear in the fourth (far-right) column of Table DC16-1. Your results will doubtless differ from mine, but hopefully not by much.

Lamp Number Two

Remove the 6.3-V lamp from its holder and replace it with a 7.5-V lamp. Using jumpers, connect the lamp to a series combination of four AA cells. As before, measure the applied voltage as shown in Fig. DC16-1A. Don't be surprised if it's different from the voltage you got across the 6.3-V lamp. Do the same thing with two AA cells in series, and then with one AA cell. Compile a table similar to the one you created for the 6.3-V lamp. The second column of Table DC16-2 shows the results I obtained.

Set your meter to measure DC amperes or milliamperes. Connect the lamp, battery, and meter together as shown in Fig. DC16-1B. Perform the experiment with four, two, and one AA cell, and enter the data into your new table. The third column of Table DC16-2 shows the values I got.

Table DC16-2   Voltages, currents, and VA power values I obtained by connecting one, two, and four AA cells in series to illuminate a 7.5-V incandescent lamp.

Number of AA Cells in Series	Volts across Lamp	Amperes through Lamp	Power as Volt-Amperes
1	1.38	0.076	0.105
2	2.65	0.110	0.292
4	5.99	0.176	1.05

Finally, as you did with the 6.3-V lamp, multiply volts by amperes to calculate VA power figures. My results are shown in the fourth column of Table DC16-2. Once again, be sure you express the potential difference in volts and the current in amperes.

Now Try This!

Repeat the voltage and current measurements with a two-diode voltage reducer in the circuit as shown in Fig. DC16-2. Connect the diodes so that their cathodes are toward the negative side of the battery, and their anodes are toward the positive side. First, use four AA cells in series. Then try two AA cells in series. Finally, conduct the tests with only one AA cell as the power source. When measuring the potential differences, connect your meter probe leads to the lamp side of the diodes, not the battery side. In each set of tests, multiply volts by amperes to get VA power figures. Are you surprised by any of the results? What can you conclude from them?

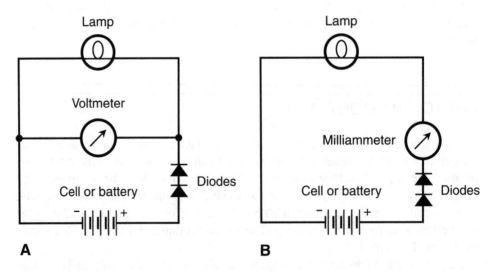

Figure DC16-2   *At A, the arrangement for measuring the voltage across a lamp with diodes in series. At B, the arrangement for measuring the current through a lamp with diodes in series. Pay attention to the diode polarity!*

DC17

Resistance as Volts per Ampere

Let's take the data from Experiment DC16 and use it to calculate the *effective resistances* of the lamp filaments under conditions of variable voltage and current. We can use Ohm's law to do this, obtaining figures technically expressed as *volts per ampere* (V/A).

What Are DC Volts per Ampere?

We can calculate the effective DC resistance of a component by measuring the voltage E across it, measuring the current I through it, and then calculating their ratio. The formula is

$$R_{V/A} = E/I$$

where E is in volts (V), I is in amperes (A), and $R_{V/A}$ is the effective resistance in volts per ampere (V/A).

When we have a resistor operating within its rated specifications, the $R_{V/A}$ figure doesn't change significantly with variations in voltage and current. In other components, particularly *transducers* designed to convert one form of energy into another (such as electrical energy into visible light), the $R_{V/A}$ figure can, and often does, change as the voltage and current vary.

Do the Arithmetic

Look again at the data you got for the 6.3-V lamp in Experiment DC16. Divide the voltage by the current to get volts per ampere. Be sure to use the correct units! The far-right column of Table DC17-1 shows my results. Do the same with the data for the 7.5-V lamp in Experiment DC16. My results appear in the far-right column of Table DC17-2.

Table DC17-1   Voltages, currents, and calculated V/A values I obtained by connecting one, two, and four AA cells in series to illuminate a 6.3-V incandescent lamp.

Number of AA Cells in Series	Volts across Lamp	Amperes through Lamp	Resistance as Volts per Ampere
1	1.25	0.093	13.4
2	2.52	0.140	18.0
4	5.85	0.250	23.4

When we calculate the $R_{V/A}$ figure for a lamp as described here, we get the effective resistance of the entire circuit. Nearly all of this resistance exists within the lamp, but all components (even the jumpers) have a little bit of internal resistance. We can keep such "stray resistances" negligible by using new batteries and making sure that our circuit connections are "solid."

Now Try This!

Measure the resistances of the lamps by disconnecting the cell or battery and using only your ohmmeter. Do the results surprise you? I got 3.2 ohms for the 6.3-V lamp and 3.4 ohms for the 7.5-V lamp—much lower resistances than when the lamps were supplied with enough current to make them glow.

All ohmmeters include internal current sources that have some inherent resistance. However, a good ohmmeter is engineered to compensate for the effects of this resistance. When I connected the ohmmeter directly to the lamps, the ohmmeter's internal current source didn't cause any visible filament glow. I concluded that the ohmmeter readings I got were good indicators of the *cold-filament resistances* of the lamps.

Draw graphs of your $R_{V/A}$ calculations as functions of the current through the lamp filaments. Create one graph for the 6.3-V lamp and another for the 7.5-V lamp. Plot amperes along the horizontal axes, and plot V/A values along the vertical

Table DC17-2   Voltages, currents, and calculated V/A values I obtained by connecting one, two, and four AA cells in series to illuminate a 7.5-V incandescent lamp.

Number of AA Cells in Series	Volts across Lamp	Amperes through Lamp	Resistance as Volts per Ampere
1	1.38	0.076	18.2
2	2.65	0.110	24.1
4	5.99	0.176	34.0

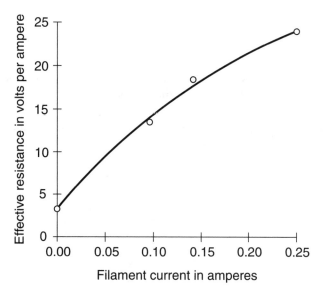

Figure DC17-1 *Effective resistance versus filament current for a DC lamp rated at 6.3 V.*

axes. For the direct ohmmeter readings, assume that the current through the filaments is zero. Figures DC17-1 and DC17-2 show the points I plotted and the graphs I created by "curve fitting."

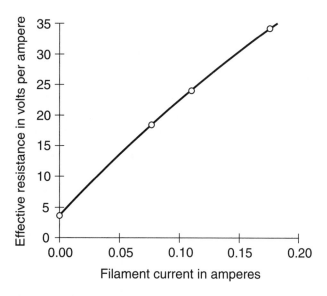

Figure DC17-2 *Effective resistance versus filament current for a DC lamp rated at 7.5 V.*

So What?

The results of these tests tell us that the internal resistance of an incandescent lamp varies, depending on how brightly the filament glows. As the brilliance of a lamp's filament increases, its temperature rises. A hot filament evidently has somewhat more resistance than a warm one, and much greater resistance than a cold one.

DC18

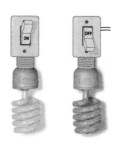

"Identical" Lamps
in Series

In a theoretically perfect series circuit, two identical incandescent lamps should share the voltage and the volt-ampere (VA) power equally. In practice, there's almost always a slight difference, because the bulb manufacturing process isn't perfect. For this experiment, you'll need two 6.3-volt (V) bulbs of the same type you used in Experiment DC16. You'll also need four AA cells connected in series, along with several jumpers.

Set Up the Circuit

Your breadboard should have at least two screw-base lamp holders. Position the board so that one holder is on the left, and another is to its right. Install a 6.3-V lamp in each socket. Connect a jumper between one terminal of the left-hand socket and one terminal of the right-hand socket. Then connect jumpers between the unused socket terminals and the battery terminals, so that the lamps are connected in series. Both lamps should glow at partial brilliance.

Call the lamp that's closer to the negative battery terminal "lamp N." Call the lamp that's closer to the positive battery terminal "lamp P." Unscrew lamp N. Lamp P will go dark. Screw lamp N back in, and then unscrew lamp P. Lamp N will go out. This phenomenon is typical of the behavior of a series circuit. If any component opens up, all the others lose power.

Short out lamp N with a jumper. Lamp N will go dark while lamp P attains full brilliance. Disconnect the jumper from lamp N, and move the jumper so that it shorts out lamp P instead. Lamp P will go dark while lamp N glows at full brilliance. Again, this behavior is typical of series circuits. If any component shorts out, all the others receive more power.

Measure the Lamp Voltages

Set your meter to read DC volts. Be sure that both lamps are glowing at partial brilliance, and then connect the voltmeter across lamp N as shown in Fig. DC18-1A. Measure the voltage and call it E_1. When I performed this test, I got a reading of 2.91 V.

Connect the meter across lamp P as shown in Fig. DC18-1B. Measure the voltage and call it E_2. I got 3.04 V, which was nearly the same (but not exactly the same) as the voltage across lamp N.

When you do these tests, you'll get results that are a little different from mine, but if things are in good working order, your results will be "in the same ball park" as mine. In any event, we can make two significant observations:

■   The voltages across the lamps are *approximately* the same.

■   The voltages across the lamps *might not* be *exactly* the same.

Although the two lamps are rated "identical," it's unlikely that they'll turn out to be perfectly matched.

When multiple components, all supposedly "identical," are put to work in a series circuit, the slightest physical discrepancy produces a significant difference in the voltages they get. This discrepancy can, and often does, render series circuits unstable. One component, getting a little more voltage, also draws more current. The result may be a "vicious circle" in which one component "hogs" more and more current, "starving" the others and putting an excessive strain on itself. In extreme situations of this sort, the "current hogging" component will burn out, destroyed by its own "greed."

Measure the Net Voltage

With your meter still set to measure DC volts, place its probe terminals across the series combination of both lamps as shown in Fig. DC18-2. Measure the voltage and call it E. In theory, you should get

$$E = E_1 + E_2$$

Based on this equation, I expected to see

$$E = 2.91 + 3.04$$
$$= 5.95 \text{ V}$$

I got a meter reading of 5.92 V.

A

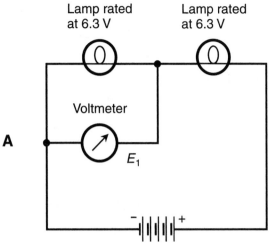

B

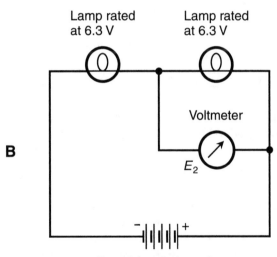

Figure DC18-1   *At A, measurement of voltage E_1 across the more negative of two "identical" lamps in series (called N, rated at 6.3 V). At B, measurement of voltage E_2 across the more positive lamp (called P, also rated at 6.3 V).*

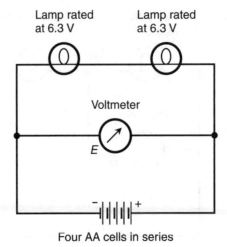

Figure DC18-2   *Measurement of voltage E across the combination of two "identical" lamps in series.*

Measure the Current

Remove the jumper wire from between the positive battery terminal and the lamps. Set your meter to measure DC milliamperes. Connect the meter where the jumper was, so that the meter is in series with the lamps as shown in Fig. DC18-3. Call the current I. When I measured I between lamp P and the battery, I got 160 milliamperes (mA), which is 0.160 ampere (A). Move the meter so as to measure I between the two lamps. Then move the meter again to measure I between lamp N and the battery. You should get the same meter reading every time, within 1 percent or less.

If the circuit current gradually decreases over time, the overall circuit resistance is rising. Such an increase in resistance can occur for either or both of the following two reasons:

■ We should expect that the circuit resistance will increase during the first few seconds the bulbs are aglow, because the filament temperatures rise as the lamps "warm up." As we've seen, the resistance of an incandescent lamp filament increases as its temperature rises. These lamps are small, so they should stabilize quickly.

■ The overall circuit resistance will increase if you leave the lamps connected to the AA cells for too long, or if any cell is weak to begin with. You might want to substitute a heavy-duty 6-V lantern battery for the set of four AA cells, and repeat all the measurements that you've done so far.

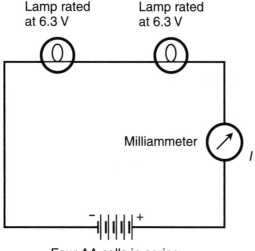

Four AA cells in series

Figure DC18-3 *Measurement of current I drawn by the combination of two "identical" lamps in series.*

Calculate the Volt-Amperes

To conclude this experiment, work out the operating VA power amounts for each lamp. Multiply the voltage across the lamp by the current through it. If P_{VA1} is the VA power consumed by lamp N, then

$$P_{VA1} = E_1I$$

When I plugged my results into this formula, I got

$$P_{VA1} = 2.91 \times 0.160$$
$$= 0.466 \text{ VA}$$

If P_{VA2} is the VA power consumed by lamp P, then

$$P_{VA2} = E_2I$$

Inputting my figures, I came up with

$$P_{VA2} = 3.04 \times 0.160$$
$$= 0.486 \text{ VA}$$

If P_{VA} is the VA power consumed by the combination of lamps, we should expect that

$$P_{VA} = EI$$

When I plugged in my measured values, I got

$$P_{VA} = 5.92 \times 0.160$$
$$= 0.947 \text{ VA}$$

In theory, the VA power consumed by the lamp combination should be equal to the sum of the volt-amperes consumed by the lamps individually. That means

$$P_{VA} = P_{VA1} + P_{VA2}$$

When I added my results of $P_{VA1} = 0.466$ VA and $P_{VA2} = 0.486$ VA, I got

$$P_{VA} = 0.466 + 0.486$$
$$= 0.952 \text{ VA}$$

The proportional error between these two results was about 0.5 percent, comfortably within acceptable limits.

DC19

Dissimilar Lamps
in Series

When mismatched incandescent lamps operate in series, they receive different voltages and consume different amounts of volt-ampere (VA) power, as this experiment demonstrates. You'll need the 6.3- and 7.5-volt (V) lamps that you used in Experiment DC16, four AA cells in series, and some jumpers.

Set Up the Circuit

You should still have the arrangement from Experiment DC18 set up. Remove the 6.3-V lamp from the socket that's closer to the positive battery terminal. Replace that lamp with one of your 7.5-V lamps, so that lamp N is rated at 6.3 V and lamp P is rated at 7.5 V. Wire the two lamps in series as you did before. Both lamps should light up about halfway, but one of them should be a little brighter than the other. When you remove either lamp, the other should go dark. If you short out either lamp, the other should shine at full, or nearly full, brilliance.

Measure the Lamp Voltages

Set your meter to read DC volts. Connect the jumpers so that the lamps are in series and are both glowing. Measure the voltage E_1 across lamp N as shown in Fig. DC19-1A. Then measure the voltage E_2 across lamp P as shown in Fig. DC19-1B. When I did these tests, I got E_1 = 2.20 V and E_2 = 3.64 V.

If you're unable to get the same Radio Shack lamps as specified in the parts list (Table DC1-1), you can use lamps from another source such as a hardware store or hobby store. This experiment will work as long as the lamps are dissimilar, and both are rated between 6 V and 12 V. You'll observe that the voltage across lamp N is significantly different from the voltage across lamp P.

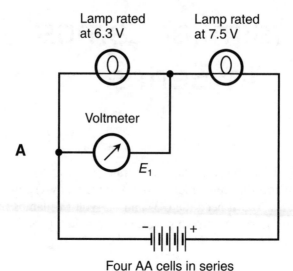

Four AA cells in series

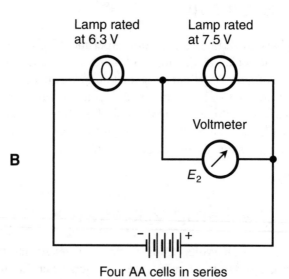

Four AA cells in series

Figure DC19-1   *At A, measurement of voltage E_1 across the more negative of two dissimilar lamps in series (called N, rated at 6.3 V). At B, measurement of voltage E_2 across the more positive lamp (called P, rated at 7.5 V).*

Measure the Net Voltage and Current

Determine the voltage E across the series combination of lamps as shown in Fig. DC19-2. In theory, the reading should be

$$E = E_1 + E_2$$

When I plugged my results into this formula, I predicted

$$E = 2.20 + 3.64$$

$$= 5.84 \text{ V}$$

My meter reading was $E = 5.85$ V, a tenth of a volt lower than the net voltage E that I measured during Experiment DC18. Apparently, my four AA cells were starting to "grow old"!

Set your meter for DC milliamperes, and connect it as shown in Fig. DC19-3 to measure the current I drawn by the series combination of lamps. I obtained a reading of 139 milliamperes (mA), equivalent to 0.139 ampere (A), considerably less than the current drawn by the pair of 6.3-V lamps in series.

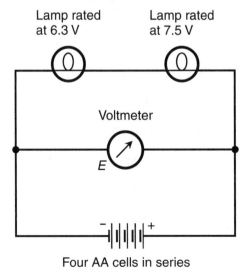

Figure DC19-2 *Measurement of voltage E across the combination of two dissimilar lamps in series.*

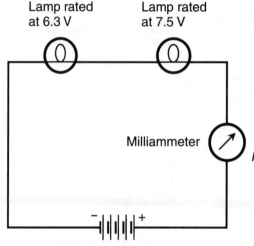

Lamp rated
at 6.3 V

Lamp rated
at 7.5 V

Milliammeter I

Four AA cells in series

Figure DC19-3   *Measurement of current I drawn by the combination of two dissimilar lamps in series.*

Calculate the Volt-Amperes

Now that you know the voltage across each lamp and the current going through the whole circuit, you can determine the VA figures for the lamps individually and together. Let P_{VA1} be the VA power consumed by lamp N, and use the formula

$$P_{VA1} = E_1 I$$

When I plugged in my results, I obtained

$$P_{VA1} = 2.20 \times 0.139$$
$$= 0.306 \text{ VA}$$

Letting P_{VA2} represent the VA power consumed by lamp P, the formula is

$$P_{VA2} = E_2 I$$

My result was

$$P_{VA2} = 3.64 \times 0.139$$
$$= 0.506 \text{ VA}$$

As before, let P_{VA} represent the VA power consumed by both lamps together. In that case, theory predicts that

$$P_{VA} = EI$$

Plugging in my experimental results gave me

$$P_{VA} = 5.85 \times 0.139$$

$$= 0.813 \text{ VA}$$

In theory, the VA power consumed by the lamp combination should be

$$P_{VA} = P_{VA1} + P_{VA2}$$

Adding my results of $P_{VA1} = 0.306$ VA and $P_{VA2} = 0.506$ VA, I got

$$P_{VA} = 0.306 + 0.506$$

$$= 0.812 \text{ VA}$$

The proportional error between my two results in this test was a small fraction of 1 percent, so I was happy with it.

DC20

"Identical" Lamps in Parallel

In an ideal parallel circuit, two identical lamps share the current and the volt-ampere (VA) power equally. In a "real-world" situation, there's usually a slight difference. For this experiment, you'll need the two 6.3-volt (V) lamps from Experiment DC18, four AA cells connected in series, and some jumpers.

Set Up the Circuit

Before starting this experiment, I replaced my four AA cells with fresh ones. The old cells showed signs of weakness. I knew that this experiment would place a significant current demand on the cells, and internal resistance could present a problem. You might want to go a step farther than I did, and use a 6-V lantern battery instead of four AA cells in series.

Position the breadboard so that one lamp socket is to the right of the other. Install a 6.3-V lamp in each socket. Connect jumpers between the socket terminals so that the lamps are wired in parallel. Then connect jumpers between the terminals of the left-hand socket and the battery. Both lamps should glow brightly. Call the lamp that's closer to the battery "lamp C." Call the lamp that's farther from the battery "lamp F."

Remove lamp C from its socket. Lamp F will remain fully lit. Screw lamp C back in, and then remove lamp F. Lamp C will still glow. This phenomenon is typical of the behavior of a parallel circuit. If any component opens up or is removed, all the other components will keep on functioning normally.

Don't connect a jumper directly across either lamp in this circuit. If you do that, you'll "short out" the battery. At best, this will drain the battery in a hurry. At worst, it could cause the battery to overheat, leak, or rupture. When any component in a parallel circuit "shorts out," all the others lose power because they're "shorted out" as well. No potential difference can exist across a component when its terminals are connected directly together!

Measure the Lamp Currents

Set your meter to read DC amperes. If you're using a digital meter, you might have to try two or more different ranges to find the one that gives meaningful readings. Be sure both lamps are fully lit. Then remove the jumper from between the positive side of lamp F and the positive side of lamp C. Replace the removed jumper with the ammeter so that the meter is in series with lamp F alone, as shown in Fig. DC20-1A. Measure the current through lamp F, and call it I_1. My meter indicated 0.24 amperes (A) when I conducted this test.

Disconnect the ammeter and replace it with a jumper. Remove the jumper from between the positive side of lamp C and the positive battery terminal. Replace that jumper with the ammeter. Move the jumper from the positive side of lamp F so that it goes directly to the positive battery terminal but *not* directly to the positive side of lamp C, as illustrated in Fig. DC20-1B. This action will ensure that the ammeter measures the current through lamp C alone. Measure this current and call it I_2. I got 0.25 A.

If your lamps are both rated at 6.3 V, and if all the meter connections were correctly done, you should be able draw two important conclusions from the results of these tests:

■ The two lamps draw *approximately* the same current.

■ The two lamps *might not* draw *exactly* the same current.

When two or more "identical" components are connected in parallel, they don't necessarily all draw the same current, because their resistances are bound to differ slightly. Nevertheless, such discrepancies don't disrupt the operation of a parallel circuit in the way they can unbalance a series circuit. Even if a "wayward" component draws somewhat more or less current than the others in a parallel connection, all of the other components will dissipate the same amount of power as they would if the "wayward" part were working properly. We'll have trouble only if the errant component's resistance is so low that it affects the ability of the battery to deliver enough current.

Measure the Total Current

Disconnect the ammeter from lamp C. Replace the meter with a jumper. Move the jumper between lamp F and the positive battery terminal, so that it's connected between lamp F and lamp C as it was originally. Both lamps should be connected

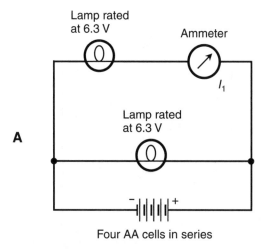

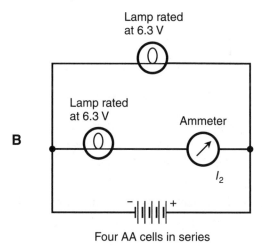

Figure DC20-1   *At A, measurement of current I_1 through the "farther-from-battery" of two "identical" lamps in parallel (called F, rated at 6.3 V). At B, measurement of current I_2 through the "closer-to-battery" lamp (called C, also rated at 6.3 V).*

in parallel, so they should both be fully illuminated. Check once again to ensure that lamp F is connected directly across lamp C.

Remove the jumper between the positive terminal of lamp C and the battery, and put the ammeter in its place as shown in Fig. DC20-2. The meter will indicate the total

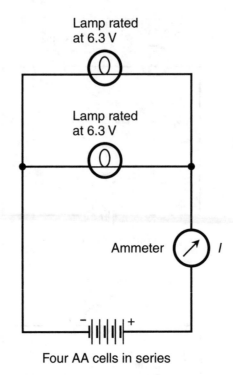

Lamp rated
at 6.3 V

Lamp rated
at 6.3 V

Ammeter   *I*

Four AA cells in series

Figure DC20-2   *Measurement of*
total current I drawn by the combination
of two "identical" lamps in parallel.

current drawn by both lamps together. Measure this current and call it *I*. In theory, the
total current should equal the sum of the individual lamp currents, as follows:

$$I = I_1 + I_2$$

Based on this formula, I expected to observe

$$I = 0.24 + 0.25$$

$$= 0.49 \text{ A}$$

My meter displayed 0.48 A.

Measure the Circuit Voltage

Replace the ammeter with a jumper, so you have the original parallel-wired
circuit operating once again. Both bulbs should glow fully. Set the meter to indicate
DC volts, and connect it directly across lamp F as shown in Fig. DC20-3. Call the

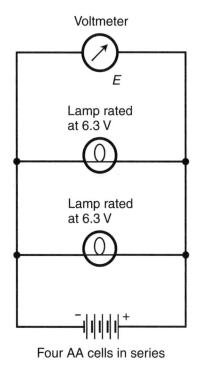

Figure DC20-3   *Measurement of voltage E across the combination of two "identical" lamps in parallel.*

voltage *E*. I got 5.50 V using four new AA cells. Measure *E* directly across lamp C, and then directly across the battery terminals. You should get the same voltage reading in each case.

If you find that the circuit voltage gradually decreases, it means that one or more of your AA cells are getting "tired," causing the internal resistance of the cell combination to rise. Try replacing the four AA cells with a new 6-V lantern battery (if you haven't been using such a battery all along!).

Calculate the Volt-Amperes

As we did in the previous two experiments, let's figure out the VA power consumed by each lamp individually, and then calculate the VA power consumed by the combination. We multiply the voltage across the lamp by the current through it to determine the VA power. If P_{VA1} is the VA power consumed by lamp F, then

$$P_{VA1} = EI_1$$

When I plugged my meter readings into this formula, I got

$$P_{VA1} = 5.50 \times 0.24$$
$$= 1.32 \text{ VA}$$

If P_{VA2} is the VA power consumed by lamp C, then

$$P_{VA2} = EI_2$$

Plugging in my figures, I calculated

$$P_{VA2} = 5.50 \times 0.25$$
$$= 1.38 \text{ VA}$$

If P_{VA} is the VA power consumed by the combination of lamps, then

$$P_{VA} = EI$$

When I plugged in my experimental readings, I got

$$P_{VA} = 5.50 \times 0.48$$
$$= 2.64 \text{ VA}$$

In theory, the VA power consumed by the lamp combination should be equal to the sum of the volt-amperes consumed by the lamps individually. Volt-amperes in parallel combine just like they do in series; they simply add. Therefore, we should have

$$P_{VA} = P_{VA1} + P_{VA2}$$

When I added my test results of $P_{VA1} = 1.32$ VA and $P_{VA2} = 1.38$ VA, I obtained

$$P_{VA} = 1.32 + 1.38$$
$$= 2.70 \text{ VA}$$

I was surprised at the difference (an error of 2 percent) between my theoretical and actual VA power figures. Apparently, the imprecision and limited resolution of my digital meter, along with *cumulative rounding errors* in my calculations, conspired in this scenario to maximize the error! This sort of thing occasionally happens in lab experiments, but it shouldn't occur very often. I hope you have better luck than I did.

DC21

Dissimilar Lamps in Parallel

When two mismatched lamps are connected in parallel, each lamp's current is inversely proportional to its resistance. The voltage is the same across both. The volt-ampere (VA) power consumed by each lamp is directly proportional to the current it carries. In this experiment, you'll analyze a "real-world" example of such a circuit. Use the 6.3- and 7.5-volt (V) lamps from Experiment DC19, four AA cells connected in series (or a 6-V lantern battery), and some jumpers.

Set Up the Circuit

Begin with the arrangement from Experiment DC20. Remove the 6.3-V lamp from the socket that's farther from the battery, and replace it with a 7.5-V lamp. Connect jumpers between the sockets so that they're wired in parallel. Then connect jumpers between the terminals of the battery and the socket that holds the 6.3-V lamp. Both lamps should glow at full or near-full brilliance. Call the 6.3-V lamp, which is closer to the battery, "lamp C." Call the 7.5-V lamp, which is farther from the battery, "lamp F."

Measure the Lamp Currents

Set your meter to read DC amperes. Be sure that both lamps are glowing. Remove the jumper from between the positive side of lamp F and the positive side of lamp C, and connect the ammeter in place of the jumper as shown in Fig. DC21-1A. The meter should indicate the current I_1 through lamp F alone. I measured 0.18 ampere (A).

Disconnect the ammeter and replace it with a jumper. Remove the jumper from between the positive side of lamp C and the positive battery terminal, and put the ammeter in the jumper's place. Move the jumper from the positive side of lamp F

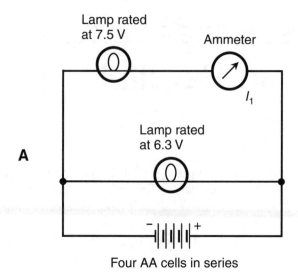

A

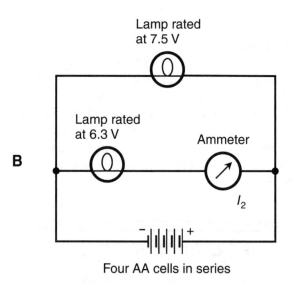

B

Figure DC21-1   *At A, measurement of current I_1 through the "farther-from-battery" of two dissimilar lamps in parallel (called F, rated at 7.5 V). At B, measurement of current I_2 through the "closer-to-battery" lamp (called C, rated at 6.3 V).*

so that it goes directly to the positive battery terminal. Your circuit should now be wired as shown in Fig. DC21-1B. The meter will display the current I_2 drawn by lamp C alone. When I measured it, I got 0.24 A.

Measure the Total Current

Disconnect the ammeter from lamp C, and put a jumper wire where the meter was. Move the jumper from between lamp F and the positive battery terminal, so that it connects lamps F and C as it did in the original circuit. Both lamps should glow. Remove the jumper from between the positive terminal of lamp C and the battery. Replace that jumper with the ammeter to create the circuit shown in Fig. DC21-2. This arrangement will tell you how much current the lamps draw in combination. Call this total current I. In theory, the following principle applies:

$$I = I_1 + I_2$$

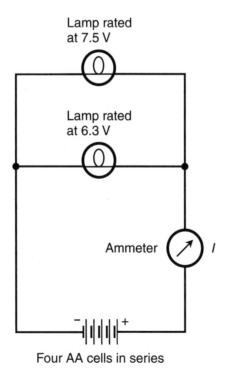

Four AA cells in series

Figure DC21-2 *Measurement of total current I drawn by the combination of two dissimilar lamps in parallel.*

Based on this formula, I expected to see

$$I = 0.18 + 0.24$$

$$= 0.42 \text{ A}$$

When I measured I directly, my meter showed 0.43 A.

Measure the Circuit Voltage

Replace the ammeter with a jumper to get the original parallel-wired circuit back once more. Both bulbs should glow fully. Set the meter for DC volts, and connect it directly across lamp F as shown in Fig. DC21-3. I measured 5.65 V in this situation. When you place the meter directly across lamp C or directly across the battery terminals, you should get the same reading as you do when the meter is across lamp F. This is the circuit voltage, which we'll call E.

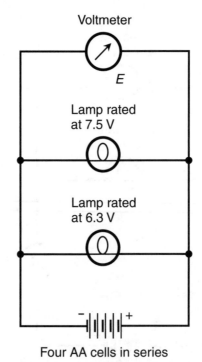

Figure DC21-3 *Measurement of voltage E across the combination of two dissimilar lamps in parallel.*

When you examine Fig. DC21-3, you'll notice that E appears across both lamps, and should be the same as the battery voltage under load. If you wonder how much the double-lamp load affects the battery voltage, try disconnecting one of the jumpers between the battery and the lamp combination so that both lamps go dark, and then measure the voltage across the battery when it's not delivering any current.

Calculate the Volt-Amperes

Let's figure out the VA power consumed by each lamp individually, and then calculate the VA power consumed by the combination. If P_{VA1} is the VA power that lamp F dissipates, then

$$P_{VA1} = EI_1$$

When I plugged my meter readings into this formula, I got

$$P_{VA1} = 5.65 \times 0.18$$
$$= 1.02 \text{ VA}$$

If P_{VA2} is the VA power consumed by lamp C, then

$$P_{VA2} = EI_2$$

Plugging in my figures, I obtained

$$P_{VA2} = 5.65 \times 0.24$$
$$= 1.36 \text{ VA}$$

If P_{VA} is the VA power consumed by the combination of lamps, then

$$P_{VA} = EI$$

When I input my results and did the calculation, I got

$$P_{VA} = 5.65 \times 0.43$$
$$= 2.43 \text{ VA}$$

In theory, the VA power consumed by the two lamps in parallel should be equal to the sum of the volt-amperes consumed by the lamps when connected one at a time. That is, we should find that

$$P_{VA} = P_{VA1} + P_{VA2}$$

When I added my separately figured values of $P_{VA1} = 1.02$ VA and $P_{VA2} = 1.36$ VA, I got

$$P_{VA} = 1.02 + 1.36$$

$$= 2.38 \text{ VA}$$

The proportional error between these two results was about 2 percent, the same as in the last part of Experiment DC20. I can't figure out why I didn't get better agreement between "theory" and "practice" in the parallel-lamp arrangements. I repeated all of the tests to double-check my work, and I got the same discrepancies again! An error of 2 percent isn't terrible, but I'd have been happier if it had been 1 percent or less.

DC22

A Compass-Based Galvanometer

In this experiment you'll observe how a current-carrying coil affects the behavior of a magnetic compass. The production of a magnetic field by an electric current is called *galvanism*. You'll need a camper's compass calibrated in degrees, 3 feet (ft) (1 meter [m]) of enamel-coated magnet wire, a sheet of fine-grain sandpaper, several resistors from your collection, six fresh AA cells, and some jumpers. You'll need your breadboard equipped with one holder for four AA cells and two holders for single AA cells. You'll also need a small mechanical punch that can put $1/4$-inch (in)-diameter holes in cardboard.

How It Works

When a magnetic compass is placed near a wire that carries DC, the compass doesn't point toward magnetic north. Instead, its needle is displaced to the east or west of north. The extent of the displacement depends on how close the compass is to the wire, and on how much current the wire carries. The direction of the displacement depends on which way the current flows through the wire.

When scientists first observed this effect, they tried different arrangements to see how much the compass needle could be displaced, and how small a current could be detected. Experimenters tried to obtain the greatest possible current-detecting sensitivity. When they wrapped the wire in a coil around the compass (Fig. DC22-1), they got a device that could indicate small currents. Once the experimenters had built this apparatus, they noticed that the extent of the needle displacement increased with increasing current.

Experimenters placed a compass on a horizontal surface so that the needle pointed toward the N on the scale (*magnetic azimuth* 0°) with no current flowing in the coil. When a source of DC voltage such as a battery was connected to the coil, the compass needle moved. As higher-voltage batteries were connected to increase the current in the coil, the compass needle deflection angle increased, but

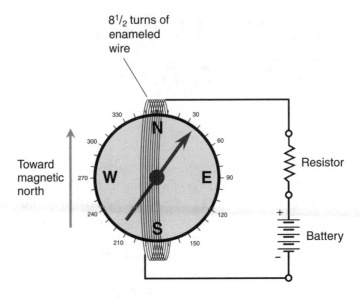

8½ turns of
enameled
wire

Toward
magnetic
north

Resistor

Battery

Figure DC22-1   *The galvanometer and associated circuitry.*
The compass must lie flat on a horizontal surface, and must be
placed so the needle points toward N (magnetic azimuth 0°)
when the battery is disconnected.

it never went past 90° either way. The meter needle might go as far as to point
toward magnetic east (azimuth 90°) or west (azimuth 270°) when the coil was connected to a massive battery. Reversing the polarity of the applied voltage reversed
the sense of the needle deflection.

Construction

Wind 8½ turns of enameled (not bare) copper wire around a magnetic compass so
that the coil turns lie along the N-S axis of the compass as shown in Fig. DC22-1. Use
small-gauge, enamel-coated magnet wire of the sort available at most Radio Shack
stores (part number 278-1345 at the time of this writing). To ensure that you get a
mechanically stable coil, paste the compass onto a small rectangular sheet of cardboard, use a punch to put small holes in the cardboard just above the N and just
below the S, and then wind the wire through the holes, passing alternately over and
under the compass. Leave a few inches of wire protruding from each end of the coil.

Use a sheet of fine sandpaper to remove approximately 1 in of enamel from
each end of the coil wire. Place the compass onto your breadboard, and wind each
bared end of the coil wire around one of the terminal nails. Figure DC22-2 shows
the layout that I used. Align the breadboard so that the compass needle points

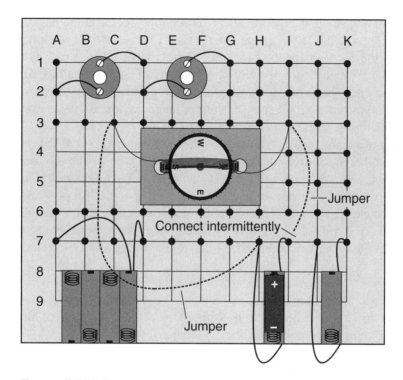

Figure DC22-2   *Suggested placement of the galvanometer on the breadboard. The jumper for the positive cell terminal is normally disconnected. This jumper should never be connected to the cell for more than a couple of seconds at a time.*

toward N on the scale (magnetic azimuth 0°). Be sure that the coil carries no current. If you've built the device properly, the needle should be aligned with the coil.

Using a jumper, connect one end of the coil to the negative terminal of a single AA cell. Connect another jumper to the other end of the coil. Then, for a couple of seconds, touch the "non-coil" end of that jumper to the positive cell terminal. The compass needle should rotate either clockwise or counterclockwise by almost 90°, so it points either slightly north of magnetic east or slightly north of magnetic west. You shouldn't leave the galvanometer connected directly to the cell for more than 2 seconds at a time, because the coil creates an almost perfect short circuit across the cell.

Testing

Take a resistor rated at 680 ohms, another rated at 470 ohms, another rated at 330 ohms, and five more rated at 220 ohms. Use the series combination of four AA cells from the past few experiments. With a jumper, connect the negative battery terminal to one end of the galvanometer coil. Choose two adjacent nails on the board as the location for the resistance to be connected in series with the galvanometer. Wrap the leads of a 680-ohm resistor around these nails. Using another jumper, connect one end of the resistor to the positive battery terminal. Switch your digital meter to a moderate DC current range. My meter has a setting for 0 to 200 milliamperes (mA), which worked well.

Firmly place one meter probe against the "non-battery" end of the resistor, and place the other meter probe against the nail where you've wound the "non-battery" end of the galvanometer coil. You should now have the circuit shown in Fig. DC22-3. The compass needle should rotate toward the east of north. If it goes west, reverse the coil connections on your breadboard to make the current flow the other way. If your digital meter displays negative current, reverse the probes so it shows positive current. Write down the readings from the digital meter and the compass azimuth scale.

Disconnect your digital meter and replace the 680-ohm resistor with one rated at 470 ohms. Repeat the current-vs.-deflection experiment. Do the same with the 330-ohm resistor, and then with the 220-ohm resistor. Keep track of all your digital meter and galvanometer readings in tabular form.

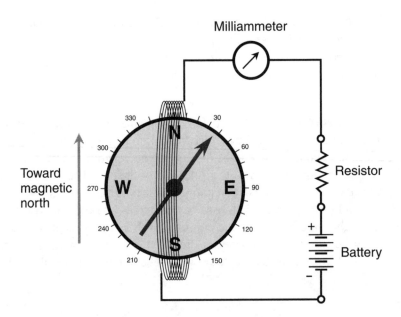

Figure DC22-3   *Arrangement for testing the galvanometer. Be sure that the compass needle points exactly toward N on the scale under no-current conditions, and deflects toward the east of N when current flows through the coil.*

Wrap a second 220-ohm resistor in parallel with the existing one, so you get 110 ohms. Repeat the measurements. Add a third 220-ohm resistor to the parallel combination, getting a net series resistance of approximately 73 ohms, and test the system again. Then add a fourth 220-ohm resistor in parallel, getting about 55 ohms; test again. Then add a fifth 220-ohm resistor to obtain a net resistance of 44 ohms, and test yet another time.

Increase the battery voltage by taking advantage of the single-cell holders on your breadboard. Place an AA cell into each holder. Wire one of them up in series with the four AA cells to get a five-cell battery, and repeat the experiment with 44 ohms of resistance. Then wire the second cell in series to get a six-cell battery, and once again, do the experiment with 44 ohms of resistance.

Calibration

When you've done all the measurements and written down all the readings from your digital current meter and galvanometer, compile a table showing the number of AA cells in the first (leftmost) column, the rated resistor values in the second column, the current levels in the third column, and the compass needle deflection angles in the fourth (rightmost) column. Table DC22-1 shows my results. Yours will doubtless differ somewhat, but they should be "in the same ball park."

Create a calibration graph by plotting the data from your version of Table DC22-1 on a coordinate grid. The horizontal axis should portray the actual current (in milliamperes), and the vertical scale should portray the compass needle deflection

Table DC22-1   Current levels and deflection angles that I obtained with AA cells and resistors in series with a compass-based galvanometer.

Number of AA Cells in Series	Resistance (ohms)	Current (milliamperes)	Deflection (degrees)
0	Infinity	0	0
4	680	9.3	7
4	470	13	12
4	330	19	19
4	220	27	28
4	110	53	40
4	73	80	50
4	55	104	54
4	44	126	60
5	44	157	64
6	44	173	66

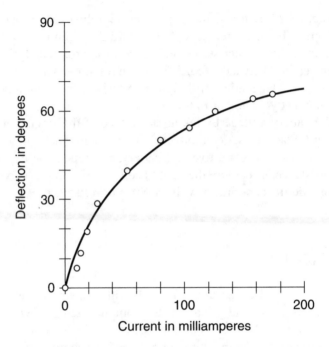

Figure DC22-4   *Compass needle deflection versus coil current for a DC galvanometer. This graph is based on my experimental results, which appear in Table DC22-1.*

angle (in degrees of magnetic azimuth). Connect the dots by curve fitting to obtain a continuous graph of deflection versus current. Figure DC22-4 shows the points and curve I got. Yours should look similar. You can now use your galvanometer, along with the calibration graph, as a crude analog milliammeter!

DC23

Solar Module in Dim Light

Solar cells are designed to operate in daylight, but they produce some voltage and current even in dim light. In this experiment, you'll measure the open-circuit voltage and the maximum deliverable current from a small solar panel at various distances from a miniature lamp. You'll need your breadboard, your digital meter, some jumpers, a yardstick graduated in inches, a small *photovoltaic* (PV) module rated at 6 volts (V) (Radio Shack part number 277-1205 or equivalent), a heavy-duty 6-V lantern battery, and the 7.5-V miniature lamp you've already used in some earlier experiments. A desk lamp or flashlight can also be helpful.

The PV Device

If you can't find the Radio Shack 6-V solar module, you can use some other PV device instead, but in that case your results will probably turn out a lot different from mine. You must use a *true PV device* designed to generate DC electricity from visible light. Don't use a *photocell*, a *photodiode*, or a *phototransistor*. They exhibit variable resistance, but they don't produce any voltage or current on their own.

You can identify a PV device by the fact that it has significant surface area (at least a couple of square centimeters), and it *looks* like a little solar panel. The Radio Shack component that I used has two wires: a black one for the negative DC pole and a red one for the positive pole. A photocell, photodiode, or phototransistor is usually much smaller than a PV device, and it *looks* like an ordinary diode or transistor.

Setting It Up

Connect the 7.5-V lamp directly to the lantern battery using two jumpers. Place the yardstick flat on your workbench with its "zero" point next to the lamp. Set your digital meter to display a low to moderate DC voltage. Using two more jumpers, connect the probe tips of your digital meter directly to the wires from the solar panel, making sure that the polarity is correct so that you'll get positive meter readings.

Your lab should be located in a room that you can *completely* darken except for the light from the 7.5-V miniature lamp, which you can use to read the digital meter. You can use a desk lamp or flashlight to aid in positioning the solar panel next to the yardstick at the correct distances from the lamp, but that light should be switched off while you measure the solar-panel output.

Measure Voltage and Current

Determine the open-circuit voltage from the solar module at distances of 1, 2, 4, 6, 8, 10, 12, 16, 20, and 24 inches (in) from the miniature lamp. Always hold the solar panel so that its surface is perpendicular to the light rays coming from the lamp, and be sure that nothing comes in between the surface of the solar panel and the lamp. Figure DC23-1 illustrates the basic geometry of the setup. The second (middle) column of Table DC23-1 shows the results I got. My digital meter gave the most meaningful readings when set for the range of 0 to 20 V.

Switch the digital meter to display small currents. My meter has a range that goes from 0 to 2000 *microamperes* (μA), and another, more sensitive range that goes from 0 to 200 μA. I used the higher-current range for levels of 10 μA or more, and the lower-current range for levels under 10 μA. The third (rightmost) column in Table DC23-1 shows my results, as obtained at distances of 1, 2, 4, 6, 8, 10, 12, 16, 20, and 24 in from the miniature lamp.

Graph the Data

The open-circuit voltage and the maximum deliverable current from the solar panel can be graphed against the distances from the lamp, and smooth curves result. However, there's a trick to this graph-plotting process! If you use straightforward *linear scales* for the coordinate axes, you'll get distorted curves. They'll be "squashed" into one corner of the coordinate grid. To make the curves fit more neatly on the grid, you can use *logarithmic scales* for the voltage and current axes.

A logarithmic graph scale differs from the more common linear graph scale. On an axis graduated in linear increments, the displacement along the axis is directly proportional to the value of the observed effect (such as voltage or current). On a logarithmic scale, however, the displacement is proportional to the *base-10 logarithm* of the observed effect, "stretching out" the curve for small values while "squashing in" the curve for larger values.

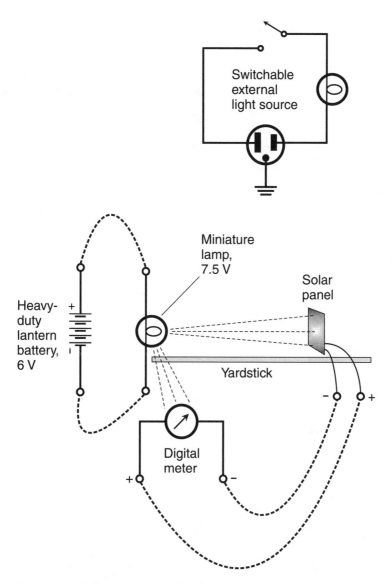

Figure DC23-1   *Arrangement for measuring solar-module output versus distance from miniature lamp. Dashed lines indicate jumper wires.*

Figure DC23-2 is a *semilogarithmic graph* of my voltage data. The term *semilogarithmic* (or *semilog*) means that one of the graph axes is linear, while the other is logarithmic. Figure DC23-3 is a semilogarithmic graph of my current data. As always, you should expect that your results, and therefore your graphs, will differ somewhat from mine.

Table DC23-1   Open-circuit voltage and maximum deliverable current from a solar module versus the distance from a miniature lamp in a darkened room.

Distance from Lamp (inches)	Voltage (volts)	Current (microamperes)
1	3.34	305
2	2.67	220
4	1.83	94
6	1.22	50
8	0.85	28
10	0.60	18
12	0.46	13
16	0.31	7.5
20	0.21	5.0
24	0.15	3.7

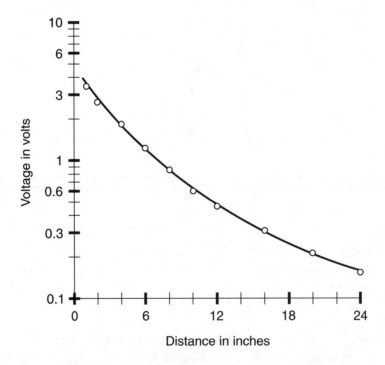

Figure DC23-2   *Open-circuit voltage from solar module versus distance from miniature lamp. The voltage scale is logarithmic.*

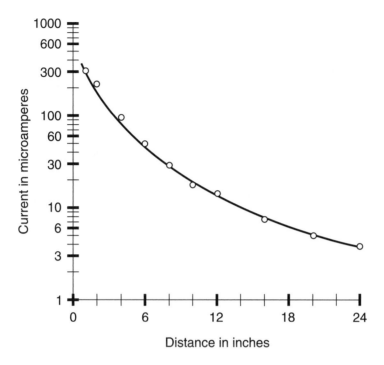

Figure DC23-3   *Maximum deliverable current from solar module versus distance from miniature lamp. The current scale is logarithmic.*

Now Try This!

Connect a resistor across the solar panel and repeat all of the previously described voltage and current measurements. Try 330 ohms first; then try a couple of other resistances such as 1 kilohm (K) and 3.3 K. When you measure voltage, connect the meter in parallel with the resistor. When you measure current, connect the meter in series with the resistor. Calculate the volt-ampere (VA) power output with the solar panel placed at distances of 1, 2, 4, 6, 8, 10, 12, 16, 20, and 24 in from the miniature lamp. Plot these VA power figures as a function of the solar panel's distance from the lamp. Use a semilogarithmic coordinate grid similar to the ones shown in Figs. DC23-2 and DC23-3.

DC24

Solar Module in
Direct Sunlight

In this experiment, you'll test a solar module to see how much voltage and current it can deliver when exposed to the midday sun. You'll need your breadboard, your digital meter, the PV module from Experiment DC23, some jumpers, and all the resistors in your collection.

Setting It Up

For best results, you should choose a cloudless day for this experiment. The sun should be high in the sky (preferably 30° or more above the horizon). In addition, sun should not be obscured, even partially, by clouds, haze, smoke, or anything else while you make your measurements. You should complete all measurements within a time frame of a few minutes. Under these idealized conditions, you can be sure that the PV module will deliver its rated output, and the intensity of the sunlight won't fluctuate enough to distort your outcomes.

If you have a long period of bad weather, or if you live where the sun doesn't shine often, you can substitute a bright indoor lamp for the sun. Any type of bulb will work—incandescent, fluorescent, compact fluorescent, or even one of those cool light-emitting-diode (LED) arrays—but you should remove the lamp shade so that the bulb is directly exposed. The solar module should be placed a foot or two away from the illuminated bulb. You might want to wear sunglasses so your eyes don't get desensitized by the brilliant light!

Measure the Voltage and Current

Carefully strip about 1 inch (in) (2.54 centimeters [cm]) of insulation from the ends of the solar module's leads. (This process can be tricky; if you're rough on the module, the wires can easily come out of it.) Place the module on the breadboard, and wrap the exposed ends of its leads several times around two adjacent terminal nails. Use the grid nails to prop up the module, so that the sun's rays will strike the surface at a right angle.

Set your digital meter to read a low to moderate DC voltage. Measure the open-circuit voltage from the module. Then set your meter to indicate DC milliamperes, and measure the current that the module delivers with no series resistance. When I performed these tests, I got 6.11 volts (V) and 55 milliamperes (mA), respectively.

Place a resistor rated at 220 ohms across the module by wrapping the resistor leads around the module-lead nails. Measure the voltage across the resistor using the arrangement shown in Fig. DC24-1A. Write down this figure. Repeat the test using load resistances of 330 ohms, 470 ohms, 680 ohms, 1 kilohm (K), 1.5 K, and 3.3 K. Write down all the voltage readings and compile them in a table. Be sure that no wayward clouds (even high, thin cirrus) creep in front of the sun at any time. My results are shown in the second column of Table DC24-1.

Go back to the 220-ohm resistance and connect your digital meter in series with it as illustrated in Fig. DC24-1B. Measure the current in DC milliamperes that flows through the resistor. Repeat the test using load resistances of 330 ohms, 470 ohms,

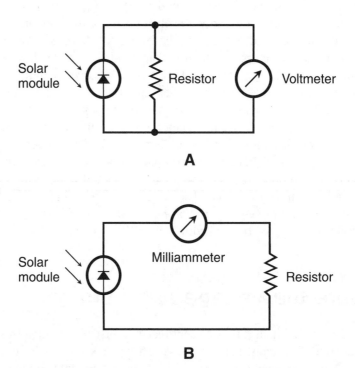

A

B

Figure DC24-1 *At A, the circuit for measuring solar-module output voltage versus load resistance. At B, the circuit for measuring solar-module output current versus load resistance.*

Table DC24-1   Voltage, current, and VmA power outputs from a solar module in direct sunlight versus the resistance placed across the module.

Resistance	Volts across Resistor	Milliamperes through Resistor	Power as Volt-Milliamperes
220 ohms	5.38	23.0	124
330 ohms	5.65	17.2	97.2
470 ohms	5.81	12.4	72.0
680 ohms	5.83	8.6	50.1
1 K	5.88	5.9	34.7
1.5 K	5.92	4.0	23.7
3.3 K	6.04	1.9	11.5

680 ohms, 1 K, 1.5 K, and 3.3 K, again making sure that all the tests are done in direct sunlight that's not muted by clouds, haze, or smoke. As before, write down all the measurements and tabulate them. My results are shown in the third column of Table DC24-1.

Calculate the Volt-Milliampere Power

Multiply the voltage across each resistance by the current through it, and write down the resulting figures in the fourth (rightmost) column of a table set up after the fashion of Table DC24-1. These figures will tell you the effective power in *volt-milliamperes* (VmA) that the solar module delivers into each resistance. A volt-milliampere is equal to exactly 0.001 volt-ampere (VA). Mathematically,

$$1 \text{ VmA} = 0.001 \text{ VA}$$

Graph the Data

Set up three coordinate grids as I've done in Figs. DC24-2 and DC24-3, with resistance on the horizontal axis. The first grid's vertical axis should show volts, the second grid's vertical axis should show milliamperes, and the third grid's vertical axis should show volt-milliamperes. You can use linear scales for all of the coordinate-grid axes. The points and curves in these figures portray the results I obtained with the Radio Shack 6-V-encapsulated solar module in direct sunlight at about 11:00 a.m. local time on a clear summer day.

Shortly after I finished making my measurements, I noticed that a high, thin layer of cirrus clouds had moved in. I didn't notice any decrease in the sun's intensity,

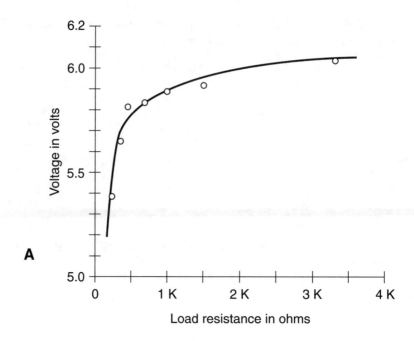

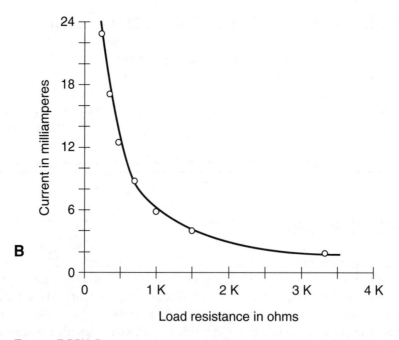

Figure DC24-2 *At A, solar-module output voltage versus load resistance in direct sunlight. At B, solar-module output current versus load resistance in direct sunlight.*

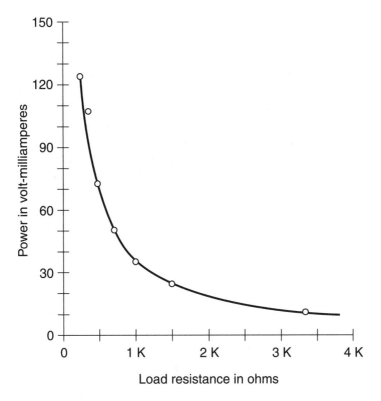

Figure DC24-3   *Graph of volt-milliampere (VmA) power from solar module versus load resistance in direct sunlight.*

but the solar module and digital meter did! All the current and voltage readings began to go down. In fact, if I hadn't been doing this experiment at the time, I probably wouldn't have noticed the clouds at all. I immediately realized that the solar module, connected directly to a digital voltmeter or a digital milliammeter, were functioning as a sensitive "light meter," technically called an *illuminometer*.

For True Experimenters Only

Figure DC24-3 suggests that if you reduce the load resistance below 220 ohms, the VmA power output of the solar module will increase. Of course, there's a limit to how much VmA power the device can deliver into any load. A bit of deductive reasoning leads us to the following conclusion: There's a certain value of load resistance for which the solar module functions at optimum *efficiency* in direct sunlight. In this context, efficiency is the ability of the module to convert radiant solar energy into electrical energy.

With some patience, you can determine the load resistance at which your solar module performs at optimum efficiency when illuminated by light of a certain constant intensity. Obtain a *linear-taper potentiometer* with a maximum resistance of 500 ohms or 1 K, and employ it as the load resistance. You should be able to find a suitable potentiometer at your local Radio Shack store. Before making your voltage and current measurements, precheck each actual resistance value with the potentiometer connected to the ohmmeter only, *and nothing else.*

After you've made enough measurements to satisfy yourself that you've "closed in" on the optimum load resistance for the module, compile the data, calculate the VmA power levels, and create a graph similar to the one in Fig. DC24-3. From this graph, you can observe the approximate optimum load resistance for the solar module. You might want to expand the horizontal axis of the graph, "stretching it out" so it indicates load resistances from 0 to 1 K, or maybe even 0 to 100 ohms.

Once you've discovered the optimum load resistance (that is, the resistance for which the VmA output power is maximum), you'll know what it takes to get the module to work at its peak efficiency *in direct midday sunlight*. What about dimmer light? If you're ambitious, repeat the efficiency tests with the solar module exposed to sunlight filtered through heavy overcast, daylight just after sunrise or before sunset, or indoor light. It's tempting to suppose that the load resistance for optimum efficiency ought to depend on the intensity of the light striking the solar module. But does it, really? There's only one way to find out!

A Photovoltaic Illuminometer

In Experiment DC23, you saw that the voltage from a solar module varies with the intensity of the light that strikes it. This property makes it possible to measure the relative intensity of visible light in any common environment. You can build a portable *photovoltaic (PV) illuminometer* using your digital meter, a resistor rated at 3.3 kilohms (K), and a solar module.

Construction

We can use a PV cell or battery to measure light intensity in either of two fundamentally different ways. We can measure the current that the device delivers through a fixed resistance, or we can measure the voltage that the device produces across a fixed resistance.

An *ideal* (theoretically perfect) ammeter, milliammeter, or microammeter exhibits zero internal resistance. In practice, the internal resistance is very low, but it's never actually zero. An ideal voltmeter, millivoltmeter, or microvoltmeter has theoretically infinite internal resistance. In practice, the internal resistance is extremely high, but not literally infinite. If we connect a fixed resistor in series with a current-measuring meter, or if we connect a fixed resistor in parallel with a voltage-measuring meter, we get a device whose reading is proportional to the square of the volt-ampere (VA) power dissipated by the resistor.

I tried to build a current-based PV illuminometer using a resistor rated at 3.3 K in series with a solar module rated at 6 volts (V) and my digital meter set to measure DC microamperes. The resulting device wasn't sensitive enough to be of use in most situations. Then I assembled a voltage-based PV illuminometer using the same resistor in parallel with the solar module and my digital meter set to measure DC millivolts or volts. This scheme, shown schematically in Fig. DC25-1, proved useful in all kinds of everyday environments, ranging from the dim light

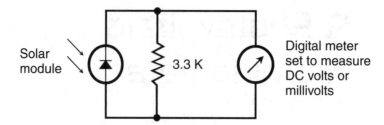

Figure DC25-1 *Schematic diagram of a voltage-based PV illuminometer.*

on a city street at night to the brilliant sunshine in an open field on a clear summer morning.

Wrap the wires from your solar module around the leads of a resistor rated at 3.3 K. Then bend the ends of the resistor leads back into U shapes, as shown in Fig. DC25-2. Insert the bent resistor leads directly into the meter input holes, compressing the Us just enough to provide snug fits and good electrical contacts. Attach the PV cell to the back of the digital meter using a short length of duct tape

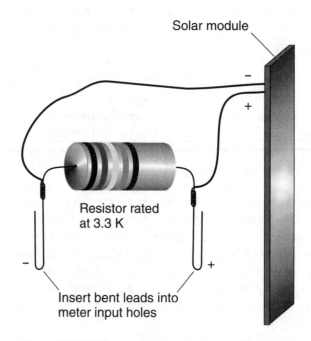

Figure DC25-2 *Arrangement for installing the solar module and parallel resistor. Be sure that the solar module polarity matches the meter polarity.*

or packaging tape rolled up so that it's "sticky-side-out." My solar module and digital test meter fit together neatly into a box measuring about 3 inches (in) wide, 4 in high, and $^3/_4$ in deep.

Testing

The fixed resistor serves an important function in this illuminometer. It forces the effective resistance of the voltmeter to remain constant, no matter what voltage range the meter is set to read. Without such a *shunt resistor*, the internal resistance of the meter would vary depending on the position of the range selector switch, creating a situation in which the behavior of the observer would affect the observed phenomenon! That sort of thing is okay in discussions of theoretical quantum physics, but we don't want it to happen in our experiments if we can possibly avoid it.

Use the PV illuminometer to measure the intensity of light in various places, and write down the meter readings in millivolts or volts. Don't let your fingers obstruct any part of the solar module when you take readings. Table DC25-1 shows readings I obtained under various conditions. Keep in mind that these readings are relative, not absolute. We aren't measuring *lumens* or *foot-candles* or any other standard unit of ambient light intensity. But we do get fairly good proportional indications of the square root of the relative light power striking the surface of the PV module.

One little "glitch" proved to be a nuisance in this experiment. I found that the digital display was difficult to read unless the illumination level remained absolutely constant. The numbers kept fluctuating, especially in bright environments. When this happened, I wrote down the best approximations I could. Don't be surprised if you experience the same trouble.

Table DC25-1   Voltages I measured across a 6-V photovotaic module connected in parallel with a 3.3-K resistor.

Environment	Output Voltage
Direct sunlight on clear morning	6.04 V
Urban landscape on clear morning	5.50 V
Office illuminated with fluorescent ceiling lamps	150 mV
Indoor swimming pool without windows	100 mV
Dimly lit office at night	41.5 mV
19-in computer display at normal reading distance, showing plain text on white background	21.5 mV
Sodium-vapor street lamp at night from 75 feet (ft) away	1.3 mV

Now Try This!

If you have a little extra money, buy an analog multimeter at Radio Shack or at your local hardware store. The meter should be capable of showing DC voltage, current, and resistance over wide ranges. Build a voltage-based PV illuminometer using the analog meter instead of the digital meter. This experiment will illustrate a major advantage of analog meter displays over digital meter displays. A digital meter usually won't work well when you want to measure phenomena that constantly change, but an analog meter with the proper amount of damping (needle-movement "sluggishness") usually will.

Part 2
Alternating Current

Your Alternating-Current Lab

You can use the same home-based laboratory arrangement for the following AC-related experiments as you did with the DC experiments. Keep your breadboard and all the components from the previous experiments. Table AC1-1 lists new items you'll need for the experiments in this section. You should be able to get nearly all of these components from Radio Shack retail stores, the Radio Shack Web site, or your local hardware store. If you can't find a certain component at any convenient retail source, you can get it (or its equivalent) from one of the mail-order outlets listed at the back of this book.

Warning! *Never touch any open connection or component while it is supplied, either directly or indirectly, with household utility power. The voltage from a "wall outlet" can drive a deadly electrical current through your body.*

Caution! *Use needle-nose pliers and rubber gloves for any wire-wrapping operations if the voltage at any exposed point might exceed 10 volts.*

Caution! *Wear safety glasses at all times as you do these experiments, whether you think you need the glasses or not.*

Table AC1-1   Components list for the AC experiments. You can find these items at retail stores near most locations in the United States. Abbreviations: ft = feet, in = inches, V = volts, W = watts, A = amperes, mA = milliamperes, PIV = peak inverse volts, and μF = microfarads.

Quantity	Store type or Radio Shack part number	Description
1	See Part 1	All components from Part 1
1	Hardware store	Multiple-outlet 15 A power strip with breaker but no surge suppressor
2	Hardware store	Three-outlet, two-wire extension cords 6 ft long
1	Hardware store	Large roll of electrical tape
2	Hardware or department store	Table or desk lamps for 117 V AC with on/off switches
2	Your house or department store	Incandescent bulbs rated at 15 W and 117 V AC
2	Your house or department store	Incandescent bulbs rated at 25 W and 117 V AC
2	Your house or department store	Incandescent bulbs rated at 40 W and 117 V AC
1	Hardware store	In-line fuse holder for 0.25-in by 1.25-in fuses
1	Hardware store	Package of 0.25-in by 1.25-in fuses rated at 1 A
1	273-1690	Power adapter, 18 or 24 V AC output at 1 A
1	272-1032	Electrolytic capacitor, 1000 μF at 35 V DC maximum
2	276-1144	Packages of two rectifier diodes rated at 3 A and 400 PIV
1	276-0563	Package of two 1NT4742A Zener diodes rated at 12 V and 21 mA
1	276-565	Package of two 1N4733A Zener diodes rated at 5.1 V and 40 mA
1	42-2454	12-ft cord with $1/8$-in mini plug on one end and spade-tongue terminals on other end

AC2

"Identical" Utility Bulbs in Series

In this experiment, you'll wire up two "identical" AC utility bulbs in series and measure the voltages across them. You'll need a pair of two-conductor, 6-foot (ft) extension cords with three outlets each, two household lamps for use with standard incandescent bulbs, two 15-watt (W) incandescent bulbs, two 25-W incandescent bulbs, two 40-W incandescent bulbs, a power strip with a 15-ampere (A) breaker (but *without* a "surge protector"), a diagonal cutter, some electrical tape, a pair of rubber gloves, and your digital multimeter.

Construct the Test Cord

Get the 6-ft extension cords from your collection of parts, and be sure neither of them is plugged into anything. Cut the wires of one cord midway between the plug and the outlets. Set aside or discard the half-cord with the male connector (that is, the pronged plug). Separate the wires on the ends of half-cord with the triple outlet, and strip 1 inch (in) of insulation from each of those wires.

Take the other 6-ft extension cord and carefully cut one—*but only one*—of its wires in the middle. Separate the free wire ends so that 6 in of wire is loose on either side of the cut. Strip 1 in of insulation from the end of each wire.

Twist-splice the two cords together as shown in Fig. AC2-1A. Wear rubber gloves while making the splices to keep the strands from puncturing your fingertips. Wrap the splices individually, and completely, with electrical tape so that *none of the copper is exposed*. You now have a two-branch "series cord." Figure AC2-1B is a schematic diagram of the arrangement.

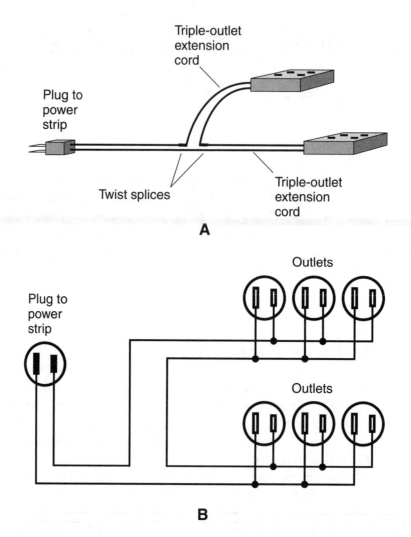

Figure AC2-1   *Extension-cord interconnection to obtain two sets of outlets in series. At A, the pictorial diagram. The twist splices must be completed and insulated before the assembly is connected to any power source. At B, the schematic diagram.*

Connect the Lamps

Be sure you're wearing shoes and the rubber gloves. Insert the series cord's male plug into the power strip, and then plug the power strip into a standard utility wall outlet. Plug a household lamp into one of the outlets at the end of each branch of the series cord. Place one lamp on your left, and call it lamp L. Place the other lamp on your right, and call it lamp R.

Install a 15-W bulb in each lamp. Ideally, both bulbs should be made by the same manufacturer (for example, General Electric or Sylvania). Switch the power strip and the lamps on. The bulbs should glow, but not as brightly as you would expect them to shine under normal conditions. Because they're connected in series, both bulbs receive less than their rated voltage.

It's reasonable to suppose that two "identical" household bulbs in series will split the voltage equally, and that the voltage across each bulb will be equal to half of the supply voltage. That's how things worked in Experiment DC18. But this is an AC circuit, not a DC circuit. Before we conclude that AC voltage behaves exactly like DC voltage in a circuit with series-connected bulbs, we must check it experimentally. Remember the skeptic's mantra: *Never assume.*

Measure the Voltages

If you wonder why triple-outlet extension cords were recommended for construction of the series cord, here's the reason: Multiple outlets facilitate easy measurement of voltages across the bulbs while minimizing the risk of electric shock. You can insert the meter probes into the slots of an unused outlet in each branch of the series cord, thereby getting an indication of the voltage across anything plugged into another outlet in that same branch.

Be sure you're wearing your rubber gloves and a good pair of shoes. Keep the 15-W bulbs in the lamps. Switch off the power strip and unplug it from the wall outlet. Set the meter to read AC volts with a maximum of at least 150 V. (My meter has a 200 V AC range, so I used it.) Insert the probes of the digital meter into the slots of one of the outlets at the end of the series cord connected to lamp L. Plug the power strip back into the wall outlet and switch the power strip on. Be sure the bulbs are glowing. Your arrangement should now be as illustrated in Fig. AC2-2A. Observe the meter reading and write it down. This is the voltage E_1 across lamp L.

Repeat all of the above-described steps for lamp R. Switch off the power strip and unplug it from the wall outlet. Insert the meter probes into one of the outlets at the end of the series cord connected to lamp R. Plug the power strip back in, switch it on, and be sure the bulbs are aglow. Your arrangement should be as shown in Fig. AC2-2B. You'll obtain a voltage E_2 that's close to, but probably not exactly the same as, E_1. My results were

$$E_1 = 60.0 \text{ V}$$

and

$$E_2 = 59.8 \text{ V}$$

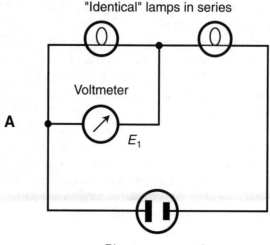

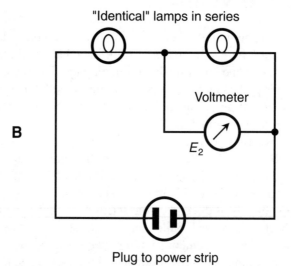

Figure AC2-2   *At A, measurement of voltage E_1 across the bulb in the left-hand of two "identical" lamps in series (called L). At B, measurement of voltage E_2 across the bulb in the right-hand lamp (called R).*

   Switch off the power strip and unplug it from the wall outlet. Insert the meter probes into one of the outlets of the *power strip* (not the series cord). Plug the power strip back in, switch it on, and be sure the bulbs are shining. Your arrange-ment should be as shown in Fig. AC2-3. Look at the meter reading and write it

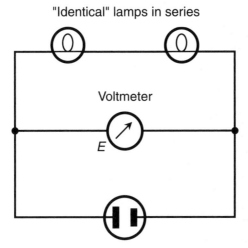

"Identical" lamps in series

Voltmeter

E

Plug to power strip

Figure AC2-3 *Measurement of voltage E across the combination of two "identical" bulbs in series.*

down. You should get a voltage that's close to the sum of the voltages across the individual lamps. When I did this test, I got a reading of $E = 120.0$ V. In theory, I expected to see

$$E = E_1 + E_2$$
$$= 60.0 + 59.8$$
$$= 119.8 \text{ V}$$

The difference between theory and practice turned out to be less than 0.2 percent, which was well within acceptable limits for experimental error.

Repeat all of the above steps using two "identical" 25-W bulbs. Then do the experiments a third time using "identical" 40-W bulbs. Your results should be similar to those you got with the 15-W bulbs. The voltages, while nearly the same, are not likely to match precisely. There are always slight differences among manufactured bulbs that are supposed to be "identical." When they're connected in series and the voltages across them are measured, even the tiniest discrepancy is exaggerated so that it becomes readily apparent.

Now Try This!

When I performed these experiments, I used bulbs of the same brand in both lamps. What do you think will happen if you try this experiment with bulbs of different brands, such as General Electric in one lamp and Sylvania in the other?

AC3

Dissimilar Utility Bulbs in Series

In this experiment, you'll measure the voltages across pairs of dissimilar AC utility bulbs connected in series. You'll need the power strip, the series cord, and the household lamps from Experiment AC2. You'll also need three incandescent bulbs: one rated at 15 watts (W), one rated at 25 W, and one rated at 40 W. Don't forget the rubber gloves and the digital meter.

15 W and 25 W

Plug the series cord into the power strip, and then plug the power strip into a wall outlet. Plug a household lamp into one of the outlets at the end of each branch of the "series cord," calling them lamps L and R as before. Install a 15-W bulb in lamp L and a 25-W bulb in lamp R. Switch the power strip and the lamps on. The bulbs should glow, but the 15-W bulb will shine more brightly than the 25-W bulb. Does this surprise you? Why do you think the "low-wattage" bulb emits more light than the "high-wattage" bulb in this arrangement?

While wearing shoes and gloves, switch off the power strip and unplug it from the wall outlet. Set your digital multimeter to read AC volts. Insert the probes into an unused outlet at the lamp L end of the cord. Plug the power strip back in, and switch it on. Be sure the bulbs are glowing. Observe the meter reading (Fig. AC3-1A) and write down the voltage E_1 across lamp L, which holds the 15-W bulb. I observed $E_1 = 90.5$ volts (V).

Switch off and unplug the power strip. Insert the meter probes into an unused outlet at the lamp R end of the cord. Plug the power strip back in, switch it on, and be sure the bulbs are aglow. Check the meter reading (Fig. AC3-1B) and write down the voltage E_2 across lamp R, which holds the 25-W bulb. I got $E_2 = 29.1$ V.

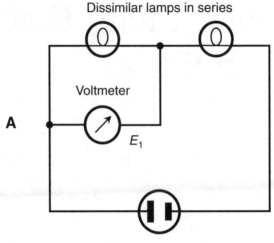

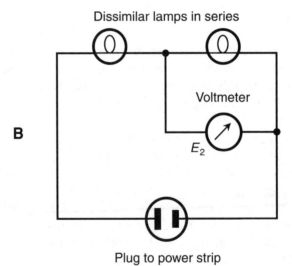

Figure AC3-1   *At A, measurement of voltage E_1 across the bulb in the left-hand of two dissimilar lamps in series (called L). At B, measurement of voltage E_2 across the bulb in the right-hand lamp (called R).*

Switch off and unplug the power strip. Insert the meter probes into an unused out-let in the power strip. Plug the strip back in, switch it on, and check the bulbs to be

Dissimilar lamps in series

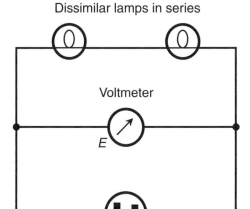

Voltmeter

E

Plug to power strip

Figure AC3-2   *Measurement of voltage E across the combination of two dissimilar bulbs in series.*

sure they're shining. Record the meter reading (Fig. AC3-2). This is the voltage E across the series combination of bulbs. I got $E = 120.2$ V. Theory predicted that I ought to have seen

$$E = E_1 + E_2$$
$$= 90.5 + 29.1$$
$$= 119.6 \text{ V}$$

The experimental error was approximately 0.5 percent, which was okay.

15 W and 40 W

Switch off and unplug everything, and replace the 25-W bulb in lamp R with a 40-W bulb. Then do all of the above-described steps again, making sure to take all the recommended safety precautions. (Don't laugh at me for repeating these warnings! An accident won't happen if you refuse to let it happen.) This time, the "higher-wattage" bulb will hardly shine at all, while the "lower-wattage" bulb will glow at nearly full brilliance. When I did these tests, I got the following results:

- The voltage E_1 across the 15-W bulb in lamp L (Fig. AC3-1A) was 106.7 V.

- The voltage E_2 across the 40-W bulb in lamp R (Fig. AC3-1B) was 13.0 V.

- The voltage E across the combination of bulbs (Fig. AC3-2) was 120.0 V.

According to theory, I expected to see

$$E = E_1 + E_2$$
$$= 106.7 + 13.0$$
$$= 119.7 \text{ V}$$

The experimental error was less than 0.3 percent.

25 W and 40 W

Once again, switch off and unplug everything. Replace the 15-W bulb in lamp L with a bulb rated at 25 W. Repeat everything, including your rigorous precautions to avoid getting an electrical shock. Here's what I found:

- The voltage E_1 across the 25-W bulb in lamp L (Fig. AC3-1A) was 86.6 V.

- The voltage E_2 across the 40-W bulb in lamp R (Fig. AC3-1B) was 32.9 V.

- The voltage E across the combination of bulbs (Fig. AC3-2) was 120.0 V.

According to theory, I expected to find that

$$E = E_1 + E_2$$
$$= 86.6 + 32.9$$
$$= 119.5 \text{ V}$$

The experimental error was roughly 0.4 percent.

So What?

The AC voltages add directly in simple circuits such as this and the one in Experiment AC2, just as if we were working with DC voltages. But that happens only because the bulbs exhibit *pure resistance* without any *reactance*. Simple appliances such as light bulbs and electric heaters have essentially pure resistances for utility AC. The AC waves in the two bulbs are always in the same phase, so they add up directly.

If the phase of the AC wave going through lamp L weren't the same as the phase of the wave passing through lamp R, the simple equation

$$E = E_1 + E_2$$

wouldn't work. If you've studied *Electronics Demystified, Teach Yourself Electricity and Electronics*, or any other book that covers AC theory in detail, I don't have to tell you why this is true. If you're curious about all this, let me make a sales pitch! Find a copy of *Teach Yourself Electricity and Electronics*, and get started on the course right away.

AC4

A Simple Utility Bulb Saver

A single diode can prolong the life of an AC household incandescent bulb. A diode-based bulb saver is easy to build. You'll need the series cord from Experiment AC3, along with the power strip, one of the household lamps, one of the 25-watt (W) bulbs, and a rectifier diode rated at 3 amperes (A) and 400 peak inverse volts (PIV). You'll also need a diagonal cutter, some electrical tape, your rubber gloves, and your digital multimeter.

Put It Together

Unplug the series cord from the power strip. Dismantle the cord by removing the tape from the connections and untwisting the wires. Install the diode in the side of the cord with the single broken wire as shown in Fig. AC4-1A. Wrap the splices individually with electrical tape so that the wire and diode leads are completely covered. Then wrap electrical tape around the entire cord, covering a span long enough to secure the diode in place and make the cord "almost like new." You now have a *half-wave rectifier* for AC. Figure AC4-1B is a schematic diagram of the device.

Plug the rectifier cord into the power strip, and then plug the power strip into a standard utility outlet. Plug a lamp into one of the outlets at the end of the cord. Install a 25-W bulb in the lamp. Switch the power strip and the lamp on. The bulb will glow at somewhat less than normal brilliance. The dimming effect occurs because half of the AC wave is blocked by the diode, so current flows during only half of the cycle. The bulb therefore operates at lower-than-normal *effective voltage* so it can "loaf."

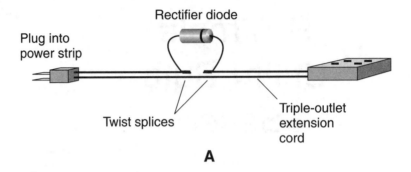

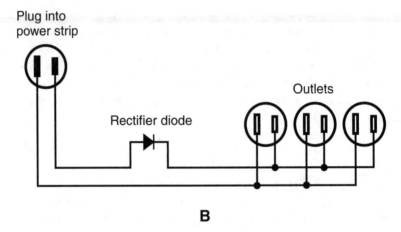

Figure AC4-1 *Drawing A is a pictorial diagram showing the construction of the rectifier cord. The twist splices must be completed, and insulated with electrical tape, before the assembly is connected to any source of power. Drawing B is the schematic circuit diagram.*

I'm using this type of rectifier cord with a desk lamp in my attic office, along with a 15-W bulb to serve as a "night light." As of this writing, that bulb has been glowing around the clock for more than four months. I don't think that it would still be shining if the bulb were directly connected to the full AC voltage, without the diode.

What's Happening?

The effective voltage produced by a fluctuating or alternating electrical source is technically known as the *root-mean-square*, or RMS, voltage. Mathematically, we determine RMS voltage by taking the square root of the mean (average) of the square of the moment-to-moment voltage (technically called the *instantaneous voltage*) over a complete wave cycle:

- First, we square the voltage at every instant in time, so that all the values become positive.

- Second, we average the *instantaneous values* (the values at each and every possible point in time) for the duration of exactly one wave cycle.

- Finally, we take the square root of the number we get at the end of the second step.

Obviously, this process cannot be done literally, because there are infinitely many time points in a wave cycle! But a computer can closely approximate the theoretical RMS value by selecting a gigantic number of evenly spaced time points during the cycle, and performing individual calculations for each point.

When we have a fairly "clean" *sine wave* of the sort that we find in household utility circuits, the RMS voltage is approximately 0.707 times the *positive peak voltage*, or about 0.354 times the *peak-to-peak voltage*. Conversely, the positive peak voltage is about 1.414 times the RMS voltage, and the peak-to-peak voltage is about 2.828 times the RMS voltage. For other waveforms such as the one that appears at the output end of our rectifier cord, these ratios are different. In this particular device, the RMS output voltage is approximately half as great it would be at the end of an ordinary cord.

When a sine-wave source of power is applied to the input of the rectifier cord shown in Fig. AC4-1B, the diode conducts during the half-cycle when the cathode is negative with respect to the anode. During the other half-cycle, the diode behaves as an open circuit. Figure AC4-2 illustrates the situation. At A, we see a graph of the AC input wave. At B, we see a graph of the rectified output wave. The diode, all by itself, forms a half-wave rectifier, which is one of the simplest possible devices for getting DC output from an AC power source.

The DC from a half-wave rectifier isn't continuous like the DC from a battery. Instead, the rectified voltage and current pulsate as time goes by. In the United States, the pulsation frequency of half-wave-rectified AC is 60 hertz (Hz), equivalent to 60 pulses per second. In some countries it's 50 Hz. That's too fast for an ordinary incandescent light bulb to follow along, so the bulb glows continuously as if it's getting DC from a battery.

Measure the Voltages

Set your digital multimeter to display DC (not AC) voltages up to 100 volts (V) or 200 V. *Be certain that the meter is set to measure voltage, not current or resistance.* One little oversight in this situation can destroy your meter. I know this from experience. Many years ago, I accidentally connected an ohmmeter directly across an active source of utility AC. You can imagine what happened!

Switch off the power strip and unplug it from the wall outlet. While wearing your rubber gloves, insert the meter probes into one of the unused outlets in the

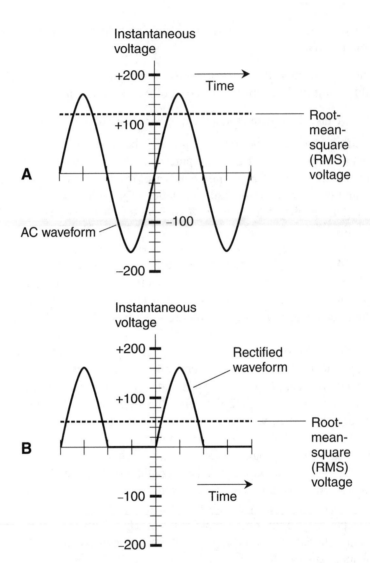

Figure AC4-2   *Half-wave rectification. At A, the AC wave-form as it appears at the power strip. At B, the output waveform as it appears at the cord outlets and across the lamp. Dashed lines indicate root-mean-square (RMS) or effective voltages.*

rectifier cord. Make certain that the positive meter lead (red wire) goes to the cathode end of the diode as shown in Fig. AC4-3A. Plug the power strip back into the wall outlet, and switch it on. Be sure that the bulb is glowing. The meter should indicate the RMS voltage as a positive value. When I conducted this test, I got a reading of +54.3 V DC.

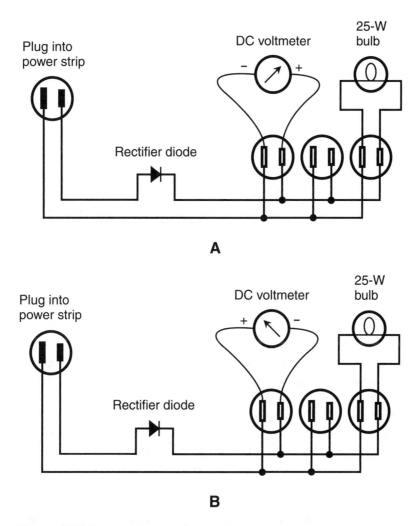

Figure AC4-3 *At A, measurement of the effective DC voltage with the meter polarity matching the rectifier output polarity. The meter shows a positive value. At B, measurement of the effective DC voltage with the meter polarity reversed. The meter displays a negative value, but the absolute voltage is the same as before.*

Switch off the power strip and unplug it from the wall outlet. Transpose the positions of the meter probes in outlet at the end of the rectifier cord. Plug the power strip back in, switch it on again, and note the meter reading. You should get the same absolute DC voltage as before, but with a negative value instead of a positive value. Figure AC4-3B shows the circuit arrangement. I got a reading of −54.3 V DC.

Switch off the power strip and unplug it yet again. Set the meter to read AC (not DC) voltages in the range of 200 V or more. Plug the power strip back in, switch it on, and note the number on the display. I got a reading of 121.7 V AC. As long as you make this measurement within a few minutes of the preceding measurements, your reading should be a good indication of the AC input voltage that the rectifier cord has received during those tests.

Now Try This!

Measure the voltages at the output of the rectifier cord again. But this time, set your digital meter for AC voltage instead of DC voltage. Again, take extra care to be sure the meter is not set to display current or resistance! Conduct tests with the meter leads connected in both directions. What does your meter say? When I did this test, I got a surprise. The meter display said 0.0 V in one direction and 121.0 V in the other! What happens when you try this test?

A Theory

I'm not sure why my meter behaved as it did under these conditions, because I didn't actually take the meter apart to see how it's constructed. But I have a theory. I think that my AC voltmeter is really a DC microammeter connected in series with a diode and a switchable set of large-value resistors. In that case, my meter has its own built-in half-wave rectifier, and it's calibrated to give a reading of RMS AC voltage by determining the pulsating DC voltage after rectification.

If I'm right, then when the diode in my meter was connected with the same polarity as the rectifier in the cord, the internal microammeter would "see" almost the same situation as it did when I measured the AC voltage directly from the power strip. The only difference would be an extra diode in series, connected in the same direction as the one in the cord. In theory, the extra diode would reduce the reading by a little less than a volt. That's what I observed.

When the meter polarity was reversed, the diode inside the meter would be connected in the opposite direction from the one in the cord. The meter's diode would cut off one half of the wave cycle while the cord's diode would cut off the other half. In theory, the two "dueling diodes" would leave the internal microammeter with zero current. That's exactly what it got.

AC5

Galvanometer with AC

For this experiment, you'll need the breadboard that you built in Part 1. You'll use a *step-down transformer* to obtain low-voltage AC from a utility outlet. Then you'll see what happens when you connect this AC source through a series resistance to a compass galvanometer.

A Note about Diagrams

In most of the remaining experiments in this section, you'll see breadboard layout diagrams, as opposed to the simple schematics used in Part 1. In general, these AC circuits are more complicated than the DC circuits in Part 1. The breadboard layout diagrams, which we might call "hybrid pictorial/schematics," will minimize the risk of wiring errors.

If you'd rather work with plain schematics and not worry about the layout particulars, feel free to convert the breadboard layout diagrams to schematic format and make up your own breadboard layouts. Translating between pictorial and schematic formats can provide you with diagram-reading and diagram-drawing practice. That's not a bad thing, even for experienced technicians and engineers.

Wire Up the Transformer

Radio Shack manufactures a step-down transformer designed for portable electronic devices. At the time of writing, this transformer, which Radio Shack calls a *power adapter*, was part number 273-1690. When plugged into a standard utility AC outlet that supplies 117 volts (V) root-mean-square (RMS), the power adapter produces either 18 or 24 V RMS AC output, depending on the position of a selector switch. For this experiment, you should set this switch to the 18 V position.

Take hold of the power adapter's output cable. That's the thin two-wire cord with the round miniature plug (not the thicker cord with the standard utility plug). Be sure that the adapter isn't connected to any source of power. Cut off the miniature plug,

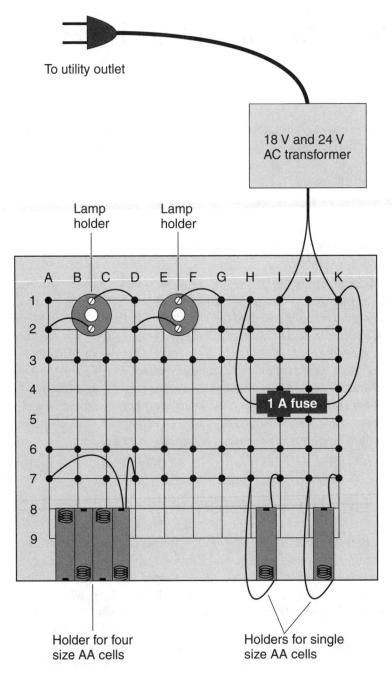

Figure AC5-1   *Breadboard arrangement for wiring the AC transformer and fuse. The low-voltage, fused AC output appears between terminals H-1 and I-1.*

separate the wires at the ends of the cord, and strip 1 inches (in) (2.5 centimeters [cm]) of insulation from each wire. Wrap one wire around terminal I-1 and the other wire around terminal K-1 on the breadboard, as shown in Fig. AC5-1.

Get an in-line fuse holder and 1-A fuse from your parts collection. Insert the fuse into the holder. Strip 1 in (2.5 cm) of insulation from the end of each wire on the fuse holder. Wrap one fuse-holder wire around breadboard terminal K-1 and the other wire around terminal H-1, as shown in Fig. AC5-1. If your breadboard is arranged in the same way as mine, you'll end up with a fused source of low-voltage AC power between terminals H-1 and I-1 as soon as you connect the transformer to a source of utility power. But don't plug the transformer in yet.

Connect the Galvanometer

Install two 1 kilohm (K), $1/2$-watt (W) resistors in parallel between terminals H-1 and H-2 on the breadboard. This parallel combination will give you a 500-ohm, 1-W resistance in series with the fused side of the transformer output. Check this resistance with your ohmmeter to be sure it's actually in the neighborhood of 500 ohms. Mine turned out to be 494 ohms, well within the 5 percent tolerance rating for both of the 1-K resistors.

Find the compass galvanometer that you built in Experiment DC22. Connect its wires to terminals H-2 and I-1 on the breadboard, thereby placing the galvanometer coil in series with the resistance at the transformer output. Once you've made these connections, you'll have the circuit arrangement shown in Fig. AC5-2. (If you built a breadboard with a design plan different from mine, refer to the plain schematic diagram of Fig. AC5-3.)

Plug the power strip into a wall outlet, and be sure it's switched off. Plug the power transformer input (the two-prong AC plug) into one of the outlets in the power strip. Place the galvanometer flat on the breadboard so that the compass is horizontal. Carefully rotate the entire breadboard/galvanometer assembly until that the compass needle points exactly toward the N, which represents magnetic azimuth 0°. Now you're ready to test the galvanometer with an AC power source. What do you think will happen?

Test the Galvanometer

Put on your rubber gloves. Once you apply power to this circuit, the highest voltage between any two exposed points on your breadboard will be 18 V RMS AC. That's not normally dangerous, but you could still get a nasty shock from it if your hands happen to be sweaty and you accidentally come into contact with the terminals.

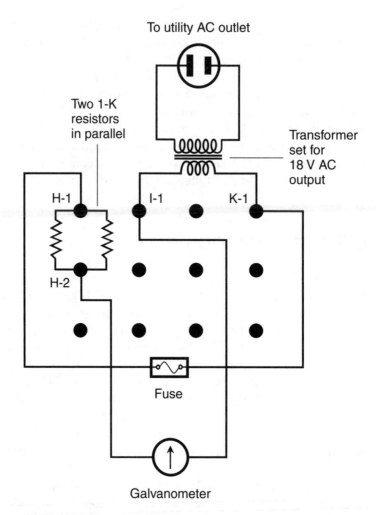

To utility AC outlet

Two 1-K
resistors
in parallel

Transformer
set for
18 V AC
output

H-1   I-1   K-1

H-2

Fuse

Galvanometer

Figure AC5-2   *Suggested breadboard arrangement for testing
a galvanometer with AC. The resistors should be rated at $^1/_2$ W,
and the transformer should be set for 18 V output. Black dots
represent breadboard terminals. Note that when two lines (representing wires) intersect in a circuit diagram without a dot, the
wires are not meant to be connected to each other at the point of
crossing.*

Switch on the power strip. Observe the galvanometer. Its needle should remain at
magnetic azimuth 0°.

Measure the AC voltage between terminals H-1 and I-1 to be sure that your power
adapter is working properly, and that the fuse isn't open. Set your digital meter for

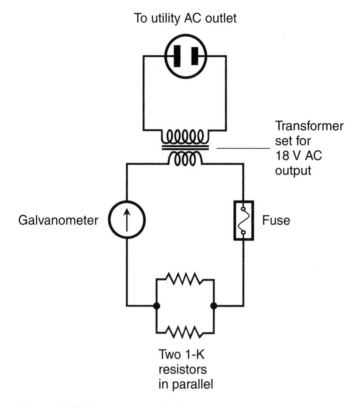

Figure AC5-3   *Simplified schematic diagram of the circuit for testing a galvanometer with AC.*

a moderate AC voltage range, hold its probe tips against the terminals, and observe the display. You should get a reading in the neighborhood of 18 V. My transformer was a little bit more energetic than its ratings suggest; I got 20.4 V.

So What?

Are you surprised by the results of this experiment? You shouldn't be. A conventional galvanometer can't measure 60-Hz utility AC, because the current reverses direction every $1/120$ of a second. The compass needle can't keep up with such rapid alternations. Before it can rotate toward the east in response to a current flowing through the coil in one direction, the current reverses and pulls the needle in a westerly direction. Then, before the needle can start to deflect toward the west, the current reverses and pulls it back east again. The alternating state of affairs continues for as long as AC flows through the coil.

AC6

Galvanometer with Rectified AC

This experiment will demonstrate how one small component—a rectifier diode—can drastically alter the behavior of an electrical circuit powered by AC. You'll need all the items you used in Experiment AC5, along with a diode rated at 400 peak inverse volts (PIV) and 3 amperes (A).

Add the Diode

Switch off the power strip, but keep the transformer plugged into it. Keep the transformer output switch to the 18-volts (V) position. Remove the galvanometer wire from terminal H-2, and move it to H-3. Wrap the cathode lead of the rectifier diode around terminal H-2. (The cathode end of the diode should be marked by a white band.) Wrap the diode's anode lead around H-3. Use a needle-nosed pliers to get the diode leads to make good contact with the terminals.

Once you've installed the rectifier diode, you should have the breadboard arrangement shown pictorially in Fig. AC6-1 and diagrammed schematically in Fig. AC6-2. The diode's cathode should be connected to one end of the parallel resistor combination. The diode's anode should be connected to one end of the galvanometer.

When power is applied to the circuit, current will flow through the galvanometer coil in only one direction. During one half of the AC cycle, electrons will be able to travel from the resistor combination through the diode and galvanometer coil to one end of the transformer's secondary winding (that is, from terminal H-2 to I-1), because the diode will act as a closed switch. But when the AC input reverses, the electrons won't be able to travel back the other way, because the diode will behave as an open switch.

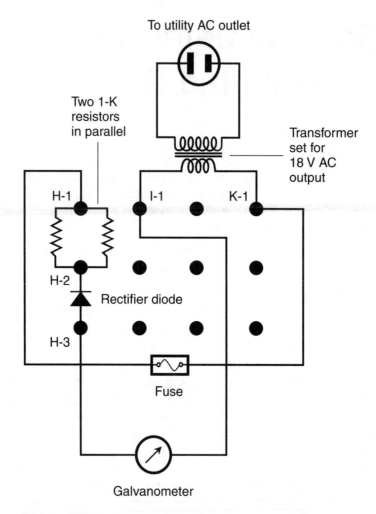

Figure AC6-1   *Suggested breadboard arrangement for testing a galvanometer with diode-rectified AC. The resistors should be rated at $1/2$ W, and the transformer should be set for 18 V output. Black dots represent breadboard terminals.*

Test the Galvanometer

Align the breadboard so that the compass needle points exactly at the N on the scale (magnetic azimuth 0°). Put on your gloves. Switch the power strip on. The compass needle should move away from the 0° point. It will deflect either eastward or westward, depending on how you have connected the galvanometer wires to the breadboard. It doesn't matter which way the needle goes. The important thing is

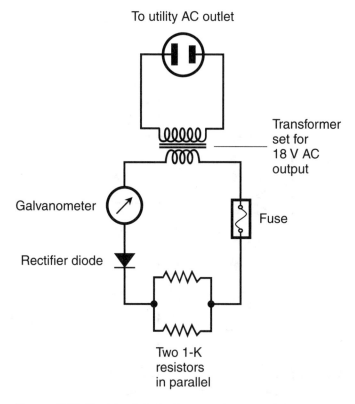

Figure AC6-2   *Simplified schematic diagram of the circuit for testing a galvanometer with diode-rectified AC.*

the extent of the compass needle's deflection. Carefully observe the number of degrees through which the needle rotates. My compass needle turned clockwise (eastward) to the 18° point on its azimuth scale.

Switch the power strip off. Remove your gloves. Using a needle-nosed pliers, carefully unwrap the leads of the rectifier diode from terminals H-2 and H-3. Reverse the polarity of the diode and wrap the leads around the same nails again, so the anode is where the cathode was before, and the cathode is where the anode was before. Put your gloves back on, switch on the power strip again, and watch the compass needle. It should rotate in the opposite sense. My compass needle turned counterclockwise to the 342° point on the azimuth scale, representing a westward rotation of 18°.

Now Try This!

Switch off the power strip. Disconnect the galvanometer lead from terminal H-3, and move it to terminal I-3. Then switch the power strip back on. Put on your gloves and set your digital meter to display DC milliamperes. Place the positive

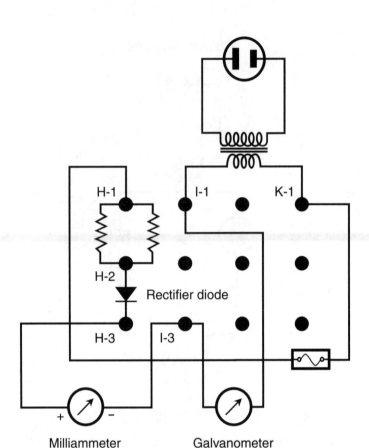

Figure AC6-3 *Interconnection of components for measuring effective rectified current through the galvanometer coil. Note the relative polarities of the diode and the milliammeter.*

meter probe tip against terminal H-3 and place the negative meter probe tip against terminal I-3, as shown in Fig. AC6-3. This arrangement will tell you the *effective current* through the galvanometer. Compare the current reading, as shown on your meter's digital display, with the galvanometer deflection in degrees. How well do these values fit onto the curve you got when you graphed the galvanometer deflection angle versus the current from a battery in Experiment DC22?

Ohm's Law with Rectified AC

In this experiment, you'll use a diode to rectify the output of your AC transformer, and see what happens under no-load conditions. Then you'll connect a resistive load and measure the effective *pulsating direct current* that the resistor draws. Finally, you'll compare your measured resistance, pulsating voltage, and pulsating current figures to see how closely they comply with Ohm's law for *pure DC* such as you get from an electrochemical battery.

Open-Circuit AC Voltages

Begin with the breadboard as it was set up for Experiment AC6. Put on your rubber gloves. Switch the power strip off and unplug the transformer. Remove the galvanometer and the resistors from the breadboard. Using a needle-nose pliers, carefully unwrap the diode leads from their terminals and remove the diode from the circuit. Plug the transformer back into the power strip, and switch the strip on.

Set the transformer for its low-voltage output [rated at 18 volts (V) RMS]. Using your digital multimeter, measure the open-circuit AC voltage across the transformer's secondary winding with the fuse in the line (Fig. AC7-1). When I conducted this test, I got 19.4 V RMS, somewhat higher than the rated voltage. Switch the transformer to the high output setting (rated at 24 V RMS). Measure the open-circuit AC voltage again. I got 26.5 V RMS.

For a moment, I wondered why my transformer outputs were higher than the rated voltages. Then I realized that the rated values are probably based on RMS AC input voltages a little bit lower than those in my house. The RMS AC output voltage from a transformer is *directly proportional* to the RMS AC input voltage. In my neighborhood, the utility voltages are usually between 120 and 125 V RMS. Some utility companies deliver lower voltages; it's common to see RMS values as low as 110 V. If my outlets had supplied 110 V RMS instead of 120 V RMS, the transformer's outputs would have been

$$19.4 \times (110/120) = 17.8 \text{ V RMS}$$

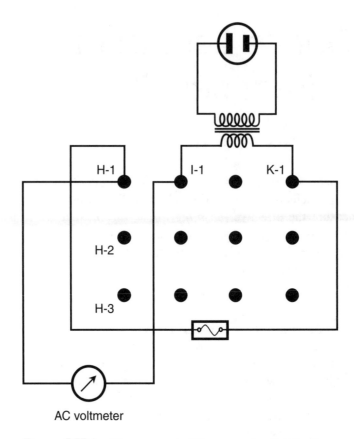

Figure AC7-1 *Measurement of the open-circuit AC voltage across the transformer's secondary winding.*

AC voltmeter

and

$$26.5 \times (110/120) = 24.3 \text{ V RMS}$$

These hypothetical values, when rounded off to the nearest volt, are exactly the same as the manufacturer's rated RMS output voltages.

Pulsating DC Voltages with Open Circuit

Wrap the leads of a 400-peak inverse volt (PIV), 3-ampere (A) rectifier diode around terminals H-2 and H-3 of your breadboard, with the cathode (banded end) at H-2. Use a jumper to connect the anode (terminal H-3) to the fused transformer secondary (terminal I-1) as shown in Fig. AC7-2. Set the transformer for its lower-voltage output. Set your digital meter for DC volts (not AC) and measure the effective open-circuit DC output voltage. Place one meter probe against terminal H-1 and the other

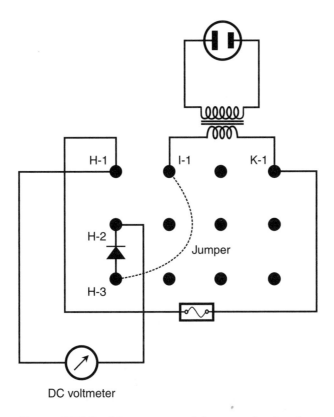

Figure AC7-2 *Measurement of the open-circuit pulsating DC output voltage from the transformer-diode combination. The jumper can serve as a "switch" for the output circuit.*

against terminal H-2. Be sure that the meter is connected to produce a positive reading, not a negative reading. Repeat the test with the transformer set at the higher-voltage output. How do the measurements of the effective open-circuit pulsating DC voltages compare with the measured RMS AC voltages?

When I performed these tests, I got meter readings of 8.42 V at the transformer's lower-voltage output setting and 11.42 V at the higher-voltage setting. These results surprised me! At first, I thought that the meter ought to "see" effective DC voltages equal to half the RMS AC voltages, because a series-connected rectifier diode "chops off" half of the AC cycle to produce the pulsating DC. However, 8.42 V is only 43.4 percent of 19.4 V, and 11.42 V is only 43.1 percent of 26.5 V.

My confusion ended when I remembered that a series-connected semiconductor diode always introduces a small *voltage drop* into a circuit. That's why the diode-based voltage reducer worked in Experiment DC15. When current passes through a semiconductor diode in the forward direction, the voltage drop across the diode can range from a fraction of a volt to approximately 1.5 V. This voltage drop occurs at every instant in time—and therefore at every point in a rectified AC wave—and that explains why my measured effective pulsating DC voltages were *less than half* of the effective AC voltages.

Pulsating DC Voltages and Currents Under Load

Set your digital multimeter to measure DC resistance. Test a 3.3-kilohm (K) resistor to determine its actual value. My resistor tested at 3.24 K. Place the resistor across the rectified transformer output as shown in Fig. AC7-3, forcing the transformer-diode combination to deliver some current.

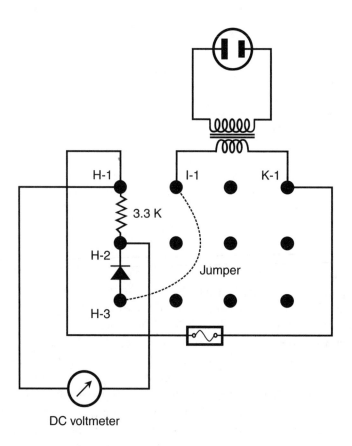

Figure AC7-3   *Measurement of the pulsating voltage from the transformer-diode combination across a resistive load.*

Set your multimeter to indicate DC voltage. Measure the voltage across the resistor at the transformer's lower-voltage output setting. When I made this measurement, I got another surprise. I expected the load to *reduce* the effective pulsating DC voltage from the transformer-diode combination. Instead, it *went up!* With no load, my circuit produced 8.42 V. With the resistive load, it produced 8.50 V.

The same thing happened when I set the transformer output switch for the higher voltage. Whereas my meter showed an effective DC voltage of 11.42 V with no load, I measured 11.72 V across the resistor. Evidently, an increase in the forward current through this particular diode reduces the voltage drop across it. Do you observe the same phenomenon when you do these tests?

Set your digital meter to indicate DC milliamperes (mA). Measure the current through the resistor as shown in Fig. AC7-4. As before, place the meter probe leads so that the digital meter readings are positive, not negative. When I did these tests, I got 2.62 mA at the transformer's lower-voltage setting and 3.62 mA at the higher-voltage setting.

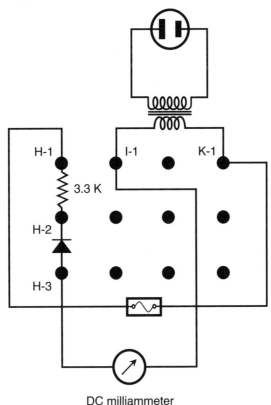

DC milliammeter

Figure AC7-4   *Measurement of the pulsating current that the transformer-diode combination drives through a resistive load.*

Letter of the Law

In Experiment DC8, we saw that Ohm's law accurately predicts the relationship among measured voltage, current, and resistance values with a pure source of DC such as we get from a battery. Let's see if "the letter of the law" holds true in the arrangement we have now. Again, the formula for Ohm's law is

$$E = IR$$

where E represents the voltage in volts, I represents the current in amperes, and R represents the resistance in ohms.

Summarize the results you got when you tested the rectifier circuit under load with the transformer set at the low-voltage output. Remember to convert the current values from milliamperes to amperes, and the resistance from kilohms to ohms. My figures were

$$E_{low} = 8.50 \text{ V}$$

$$I_{low} = 0.00262 \text{ A}$$

$$R_{low} = 3240 \text{ ohms}$$

According to the formula, I should have seen a voltage across the resistor of

$$E_{low} = I_{low}R_{low}$$

$$= 0.00262 \times 3240$$

$$= 8.49 \text{ V}$$

The discrepancy between the theoretical and actual voltages was only 0.01 V, representing a small fraction of 1 percent. With the transformer output switch set to the high-voltage position, my figures were

$$E_{high} = 11.72 \text{ V}$$

$$I_{high} = 0.00362 \text{ A}$$

$$R_{high} = 3240 \text{ ohms}$$

According to the formula, I should have seen a voltage across the resistor of

$$E_{high} = I_{high}R_{high}$$

$$= 0.00362 \times 3240$$

$$= 11.73 \text{ V}$$

Again, theory and practice agreed almost perfectly.

Now Try This!

Conduct the above-described measurements and calculations with a load resistance of 1.5 K. Then do it all a third time, using two 1.5-K resistors in parallel as the load. What can you conclude about the diode's forward voltage drop as the load resistance decreases, causing the current to increase?

AC8

A Simple Ripple Filter

The pulsating output of a half-wave rectifier won't operate most electronic devices properly unless the voltage and current pulsations (also called *ripples*) are "smoothed out" by a *ripple filter*. In this experiment, you'll build a simple ripple filter by adding an *electrolytic capacitor* having a value of 1000 *microfarads* (µF) and rated at 35 *working volts DC* (WVDC) to the arrangement from Experiment AC7.

Install the Capacitor

A *filter capacitor* works by "trying" to maintain a pulsating DC voltage at its peak level. It takes a lot of capacitance to get rid of the ripples in the output of a half-wave power supply such as the one you built in Experiments AC6 and AC7. Electrolytic capacitors provide high capacitance per unit volume, so they're ideal for use in power-supply filters. Such a capacitor is *polarized*, meaning that it must be connected in the right direction in order to function.

Start with the arrangement shown in Fig. AC7-3 on page 184. Switch the power strip off. Remove the resistor. Set the transformer switch to the lower-voltage position [rated at 18 volts (V) RMS]. Put on your rubber gloves, and wear a pair of safety glasses to protect your eyes. Do you feel as if you're in a high-profile chemistry lab? Well, you are! Electrolytic capacitors work by chemical action. A mistake with the "mix" can cause the chemicals to misbehave. If you install an electrolytic capacitor "backward" and then apply voltage to it, the little thing will explode like a firecracker. (I'm not joking. I've seen it happen.)

Switch the power strip on, but don't install the capacitor yet. Use your DC voltmeter to verify the polarity at terminals H-1 and H-2. Terminal H-1 should be negative, and terminal H-2 should be positive. Switch the power strip back off.

Now check the polarity of the electrolytic capacitor. My unit (Radio Shack part number 272-1032) has a black stripe with a minus sign on one side of the case, indicating that the lead next to the black stripe should go to the more negative voltage point in the circuit.

> **Warning!** *If you can't find a polarity marking on your electrolytic capacitor, return it to the place where you bought it or else throw it away. Obtain a capacitor that clearly shows the proper working polarity. If you can't find one, don't do this experiment!*

In the arrangement of Fig. AC8-1, the negative voltage appears at terminal H-1, which is connected to the fuse. The capacitor's positive lead must go to terminal H-2, which is connected to the cathode (banded end) of the diode. Wrap the capacitor wires around the terminals. When you're sure everything is in place and you've got the polarity right, switch the power strip on.

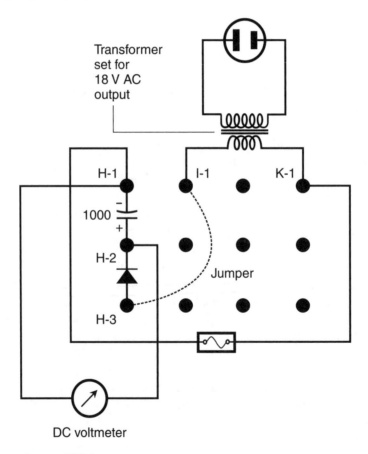

Figure AC8-1 *Addition of a filter capacitor to the half-wave rectifier circuit, and measurement of the open-circuit filtered DC voltage across the capacitor. The value (1000) is stated in microfarads (μF). Pay careful attention to the diode and capacitor polarities! The jumper can serve as a "switch" for the output circuit.*

Open-Circuit Voltage

Set your digital meter to display a relatively high DC voltage. My meter has a set-ting for 0 to 200 V DC, which worked fine for these tests. Place the meter's negative (black) probe tip against terminal H-1, and place the positive (red) probe tip against terminal H-2 as shown in Fig. AC8-1. Be sure that the jumper wire is connected between terminals H-3 and I-1. The meter should indicate a positive voltage. Are you surprised at the value? It should be considerably higher than it was in Experi-ment AC7. Whereas my half-wave rectifier circuit produced 8.42 V of pulsating DC without filtering and without any load, the addition of the filter capacitor increased the no-load output to 28.5 V DC.

When there's no load, or when the load resistance is high, the voltage across the capacitor holds steady near the *peak* rectifier output voltage as shown in Fig. AC8-2. Therefore, the effective voltage across the capacitor is a lot higher than the RMS value of the half-wave pulsating voltage that charges it. (The effective voltage doesn't increase if you place the capacitor across a source of DC that's pure to begin with, such as a lantern battery. Try it and see! As always, be sure to get the capacitor polar-ity correct.) This dramatic voltage increase is the reason why, whenever you build a DC power supply filter, you should use a capacitor that's rated to handle several times the rectifier's effective output voltage.

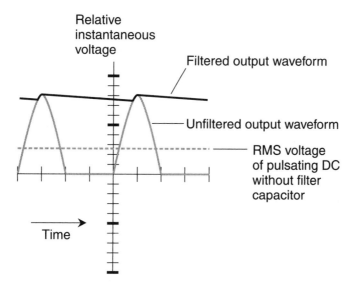

Figure AC8-2   *A filter capacitor charges up to the peak rectifier output voltage with each DC pulse. If the load resis-tance is high or infinite, the capacitor holds most or all of its charge, so the filtered output voltage is much higher than the RMS output voltage in the absence of the capacitor.*

Voltage under Load

Remove power from the circuit by switching off the power strip. Place a jumper directly across the electrolytic capacitor, and leave it connected for a few seconds. This action will discharge the capacitor. Remove the discharging jumper and then install a resistor rated at 3.3 kilohms (K) and $1/2$ watt (W) between terminals H-1 and I-2 as shown in Fig. AC8-3. Check the resistor's value with your ohmmeter to be sure that its actual value is "in the ball park." Mine turned out to be 3.24 K. Place a short length of bare copper wire between terminals H-2 and I-2, thereby connecting the resistor in parallel with the filter capacitor.

Switch the power strip back on. The power supply is now working into a resistive load. Set your meter to read DC volts, and measure the voltage across the load as shown in Fig. AC8-3. Has it changed from its value when there was no load? I measured 27.7 V DC with the 3.24-K load, a decrease of 0.8 V, indicating that the capacitor wasn't able to charge up quite as much as before. This

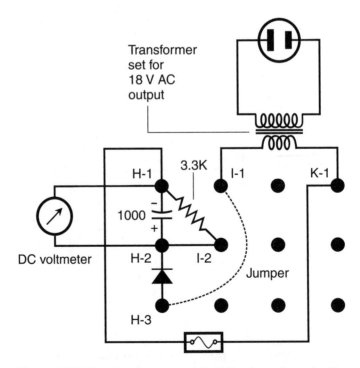

Figure AC8-3 *Measurement of the DC voltage from the filtered half-wave power supply with a resistive load. Note the shorting wire between terminals H-2 and I-2.*

result didn't surprise me, although I had imagined that the voltage would go down more.

Current through Load

Once again, switch the power strip off to shut down the circuit. Discharge the capacitor by momentarily shorting it out with a jumper wire. Set the digital meter to indicate DC milliamperes. Remove the shorting wire from between terminals H-2 and I-2. Switch the power strip on, and then place the meter probe tips against

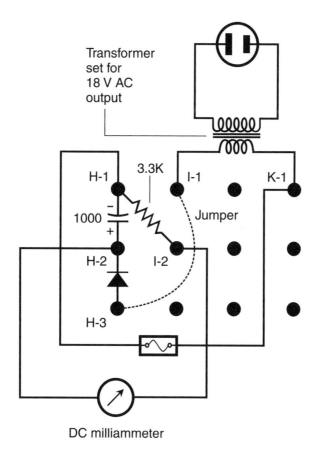

DC milliammeter

Figure AC8-4 *Measurement of the current that the filtered half-wave power supply drives through a resistive load. Note that the shorting wire between terminals H-2 and I-2 has been removed.*

terminals H-2 and I-2 as shown in Fig. AC8-4. The meter should indicate a few milliamperes (mA). When I did this test, I got 8.50 mA.

Now check the voltage, current, and resistance values to see how well they comply with Ohm's law. Remember to convert the resistance to ohms and the current to amperes. Use the formula

$$I = E/R$$

where I represents the current in amperes, E represents the voltage in volts, and R represents the resistance in ohms. My measured values were

$$E_{load} = 27.7 \text{ V}$$
$$R_{load} = 3240 \text{ ohms}$$
$$I_{load} = 0.00850 \text{ A}$$

According to the formula, I should have seen a current through the resistor of

$$I_{load} = E_{load}/R_{load}$$
$$= 27.7/3240$$
$$= 0.00855 \text{ A}$$

The discrepancy between my theoretical and actual currents amounted to about 0.6 percent. That was all right.

Now Try This!

Conduct the above described voltage measurement with two 3.3-K resistors in parallel as the load. Then do it using three 3.3-K resistors in parallel as the load. Then try it yet again with four 3.3-K resistors in parallel. What can you conclude about the power supply's output voltage as the load resistance decreases, thereby forcing the system to deliver more current?

AC9

Rectifier and Battery

In this experiment, you'll place a battery in series with the unfiltered output of a half-wave rectifier. The following tests might at first seem like a "repeat" of Experiment DC2, in which you connected cells and batteries in series—but wait until you see the results! You'll need all the components from the previous few experiments, along with a fresh lantern battery rated at 6 volts (V).

Set It Up

Switch off the power strip. Then rearrange the breadboard components so that they're interconnected as shown in Fig. AC9-1. Pay special attention to the orientation of the rectifier diode. Its cathode (banded end) should go to terminal H-1, and its anode should go to I-1. The "non-transformer" lead from the fuse holder should go to H-3.

Rectifier Alone

Put on your rubber gloves. Set the transformer for its lower-voltage output. Plug the transformer into the power strip, and switch the strip on. Set your digital meter to measure DC volts, and hold its probe tips firmly against terminals H-1 and H-3. The negative meter probe tip should go to H-3, and the positive meter probe tip should go to H-1.

Observe the meter display. Compare the reading with what you got when you did Experiment AC7 and the transformer was set for its lower-voltage output. Is there a difference? If so, it can be explained by the fact that you're doing this experiment at a different time (and most likely on a different day) than you did Experiment AC7. Your household utility voltage is probably higher or lower now than it was then. When I did Experiment AC7, I observed 8.42 V on the meter. This time, I saw 9.05 V—an increase of more than 7 percent!

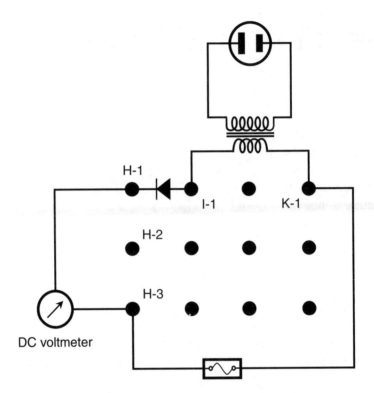

Figure AC9-1   *Measurement of the open-circuit pulsating DC output voltage from the transformer-diode combination alone. This is the same circuit that you built according to Fig. AC7-2, but the components are arranged differently to accommodate future tests.*

Switch the transformer to its higher-voltage output and repeat the measurement of the rectified, pulsating DC voltage. In this situation, my meter said 12.33 V, compared with 11.42 V in Experiment AC7, representing an increase of about 8 percent. I'm glad that I repeated these tests to guarantee that I got good initial data for the experiments to follow. (Do you see a lesson here?)

Battery Aiding Rectifier

Before you connect the battery to anything, measure its pure DC output voltage. My battery tested at 6.44 V. Once you've tested the battery and written down the results, put on your gloves, switch the transformer back to its lower-voltage output, and connect the battery in series with the rectifier output as shown in Fig. AC9-2. The

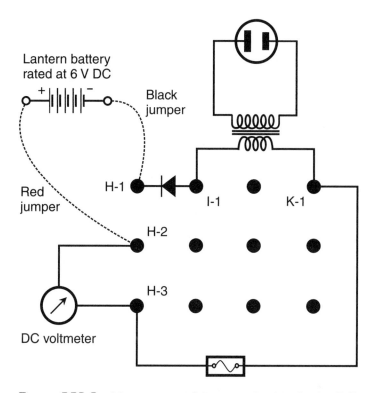

Figure AC9-2 *Measurement of the open-circuit pulsating DC output voltage from the transformer-diode-battery combination with the battery connected "forward" so its voltage adds to the rectified voltage.*

negative battery terminal should go to terminal H-1 on the breadboard. Use a black jumper wire to remind yourself that you're working with the negative side of the battery. Use a red jumper to connect the positive battery terminal to H-2. These connections will cause the battery to work with (or *aid*) the rectifier.

Keep your digital meter set to display DC voltage. Place the negative meter probe tip in contact with H-3, and place the positive meter probe tip in contact with H-2. Note the meter reading. Are you surprised? I was! According to the rule for combining DC voltages in series, I expected (until I gave the matter more serious thought) to see a net voltage, E_{net}, equal to the sum of the rectifier output voltage E_{rect} and the battery voltage E_{batt}; that is,

$$E_{net} = E_{rect} + E_{batt}$$

$$= 9.05 + 6.44$$

$$= 15.49 \text{ V}$$

My rectifier-battery arrangement didn't obey this "law." My meter indicated 12.14 V. Instead of adding 6.44 V to the root-mean-square (RMS) output of the rectifier, the battery contributed only an additional 3.09 V. Evidently, a transformer-rectifier combination doesn't behave like an electrochemical battery when connected in series with another source of voltage.

Now set the transformer to its higher-voltage output and repeat the above test. According to my theoretical "law" for addition of voltages from electrochemical cells and batteries, I should have observed

$$E_{net} = E_{rect} + E_{batt}$$

$$= 12.33 + 6.44$$

$$= 18.77 \text{ V}$$

As with the lower-pulsating DC voltage, my meter registered a lower voltage than pure theory would suggest: only 15.30 V. In this situation, my series-connected battery boosted the pulsating DC output by 2.97 V.

Battery Bucking Rectifier

Switch off the power strip to shut down the circuit. Switch the transformer back to its lower-voltage setting. Disconnect the battery jumpers from the breadboard terminals. Then re-connect them in the opposite sense from before: red (the positive side of the battery) to H-1 and black (the negative side of the battery) to H-2. Your circuit arrangement should now conform to Fig. AC9-3. The battery will work against (or *buck*) the rectifier output.

Once you're satisfied that everything is in order, be sure that you're wearing your gloves, and switch on the power strip. With your digital meter set for DC volts, place the negative meter probe tip in contact with H-3, and place the positive meter probe tip in contact with H-2. According to the DC-voltage combination rule for series circuits, my meter should have shown a net voltage, E_{net}, equal to the difference between the rectifier output voltage E_{rect} and the battery voltage E_{batt}. Because the rectifier voltage was the greater of the two, I subtracted the battery voltage from the rectifier voltage to get a theoretical result of

$$E_{net} = E_{rect} - E_{batt}$$

$$= 9.05 - 6.44$$

$$= 2.61 \text{ V}$$

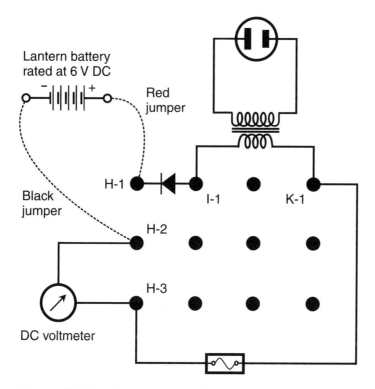

Figure AC9-3   *Measurement of the open-circuit pulsating DC output voltage from the transformer-diode-battery combination with the battery connected "backward" so its voltage subtracts from the rectified voltage.*

My meter display indicated 5.62 V. Having seen what happened in the previous test, I wasn't surprised. Instead of taking away 6.44 V from the RMS output of the rectifier, the battery caused a voltage reduction of only 3.43 V.

Switch the transformer to its higher-voltage setting and measure the rectifier-battery output voltage again. The battery-combination formula would suggest that I ought to have seen

$$E_{net} = E_{rect} + E_{batt}$$

$$= 12.33 - 6.44$$

$$= 5.89 \text{ V}$$

My meter displayed 9.01 V, indicating that the series-connected battery reduced the effective pulsating DC output by only 3.32 V.

What's Happening?

The pure DC voltages from electrochemical batteries and cells add and subtract straightaway when we connect them in series, as we saw in Experiment DC2. But things aren't so simple when one of the sources pulsates. We can't directly add a pure DC voltage to the RMS voltage from a fluctuating source and expect to get the correct RMS voltage of the series combination. Instead, we must add the voltages *at each instant in time*, point-by-point, then determine the resulting voltage waveform, and finally calculate the RMS voltage of that waveform. My digital meter carried out that process electronically as I measured the voltages with the battery in series with the rectifier output. The mathematical details of the calculation process are beyond the scope of this discussion, but I worked out an approximation. I got theoretical results that agreed closely with the outcomes of these experiments.

Now Try This!

Conduct the above-described measurements with a 3.3-kilohm (K) resistor to provide a load for the battery-rectifier combination. Connect the resistor between H-2 and H-3 in the arrangements of Figs. AC9-2 and AC9-3. Try the experiment again using two, three, and finally four 3.3-K resistors in parallel as the load. How do the voltages across these loads compare with the no-load voltages?

Rectifier/Filter and Battery

You've seen that the voltages from an unfiltered half-wave rectifier and an electrochemical battery don't add or subtract "neatly" in series. In this experiment, you'll discover how the voltages add when the rectifier ripple is eliminated by a filter capacitor, so that both sources produce pure DC. Start with the arrangement from Experiment AC9. Recover the filter capacitor you used in Experiment AC8.

Set It Up

Switch off the power strip. Add the 1000-microfarad (µF), 35-working-volts-DC (WVDC) filter capacitor to the circuit of Experiment AC9, obtaining the component arrangement shown in Fig. AC10-1. Be careful with the capacitor polarity, as you were in Experiment AC8. The positive lead of the capacitor should go to terminal H-1, which also goes to the cathode of the rectifier diode. The capacitor's negative lead should go to I-2. Cut a 3-inch (in) length of fine, bare copper wire and connect it between terminals I-2 and H-3.

Rectifier and Filter Alone

Put on your gloves and glasses. Set the transformer for the lower-voltage output, and plug it into the power strip. Switch on the power strip. With your digital meter set to indicate DC volts, measure the voltage of your lantern battery again, just to be sure that it hasn't weakened. My battery maintained its previous output of 6.44 volts (V).

Now hold the meter probe tips between terminals H-1 and H-3, as shown in Fig. AC10-1. The negative meter probe tip should go to H-3, and the positive probe tip should go to H-1. Compare the voltage reading with what you got in

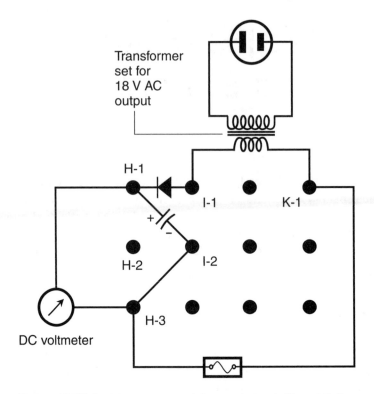

Figure AC10-1   *Measurement of the open-circuit filtered DC output voltage from the transformer-diode-capacitor combination alone. This is the same circuit that you built according to Fig. AC8-1, but the components are arranged differently to accommodate future tests.*

Experiment AC8 with the transformer set for its lower-voltage output. The filtered voltage should be nearly the same as it was then. The circuit here is identical to the one that you built in Experiment AC8, but the output voltage might differ because of changes that have taken place in your household utility voltage since that time. I observed a filtered, open-circuit output of 29.1 V.

Battery Aiding Rectifier/Filter

Connect the battery in series with the filtered, rectified power-supply output. The negative battery terminal should go to H-1 and the positive battery terminal should go to H-2, as illustrated in Fig. AC10-2. With the meter still set for DC

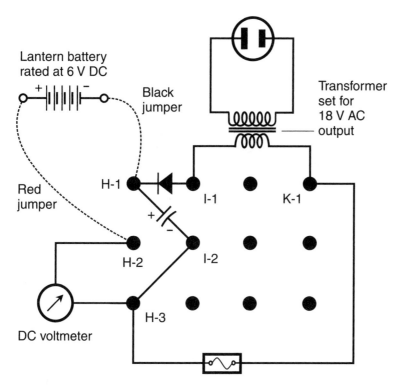

Figure AC10-2   *Measurement of the open-circuit filtered DC output voltage from the transformer-diode-capacitor-battery combination with the battery connected "forward" so its voltage adds to the power-supply voltage.*

volts, hold its negative probe tip against H-3 and its positive probe tip against H-2. According to the rule for combining pure DC voltages in series, I predicted that the net voltage, E_{net}, would be equal to the sum of the filtered output voltage E_{filt} and the battery voltage E_{batt}, as follows:

$$E_{net} = E_{filt} + E_{batt}$$

$$= 29.1 + 6.44$$

$$= 35.5 \text{ V}$$

rounded to the nearest tenth of a volt. My meter showed 35.6 V. The complicating factors in Experiment AC9 were gone! The installation of the filter capacitor brought the rectifier circuit into compliance with "conventional DC wisdom."

Battery Bucking Rectifier/Filter

Switch off the power strip to shut down the circuit. Keep the transformer at its lower-voltage setting. Disconnect the battery jumpers from the breadboard, and then re-connect them in reverse so that you get the circuit shown in Fig. AC10-3. This arrangement will force the battery to buck the filtered power-supply output.

Be sure that you have your gloves and glasses on, and then activate the power strip. Place the negative probe tip of your digital meter against H-3. Place the positive probe tip in contact with H-2. I expected to see a net voltage, E_{net}, equal to the difference between the filtered power-supply output voltage E_{filt} and the battery

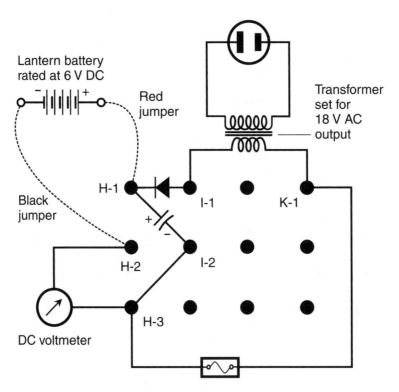

Figure AC10-3 *Measurement of the open-circuit filtered DC output voltage from the transformer-diode-capacitor-battery combination with the battery connected "backward" so its voltage subtracts from the power-supply voltage.*

voltage E_{batt}. Because the power-supply voltage was greater than the battery voltage, I did the following calculation:

$$E_{net} = E_{filt} - E_{batt}$$
$$= 29.1 - 6.44$$
$$= 22.7 \text{ V}$$

My meter indicated 22.8 V. This outcome, along with the result I got when the battery aided the power supply, convinced me that the 1000-μF capacitance got rid of essentially all the ripple in the rectifier's output under no-load conditions.

Now Try This!

Conduct the above-described tests with a 3.3-kilohm (K) resistor to provide a load for the transformer-diode-capacitor-battery combination. Use only the lower-voltage transformer setting to avoid exceeding the filter capacitor's working voltage. Connect the resistor between H-2 and H-3 in the circuits of Figs. AC10-2 and AC10-3. Then repeat the tests using two, three, and finally four 3.3-K resistors in parallel to serve as the load. Compare the voltages across these loads with the open-circuit voltages.

AC11

Rectifier and Battery under Load

In this experiment, you'll verify Ohm's law with a battery connected in series with the unfiltered output of a half-wave rectifier, driving current through a resistance. You'll need all the components from the previous few experiments, along with a resistor rated at 3.3 kilohms (K) and $1/2$ watt (W).

Set It Up

Switch off the power strip. Rearrange the breadboard components so that they're interconnected as shown in Fig. AC11-1. Be sure that the diode is connected with the cathode to terminal H-1 and the anode to I-1. Move the "non-transformer" wire of the fuse holder to I-3. Place a 3.3-K resistor between H-3 and I-3. Before applying power to the circuit, use your ohmmeter to measure the actual value of the resistor. Mine tested at 3.25 K.

Rectifier Alone

Put on your rubber gloves. Set the transformer for its lower-voltage output. Plug the transformer into the power strip, and then switch on the strip. Set your digital meter to measure DC milliamperes. Hold the positive meter probe tip against terminal H-1, and hold the negative meter probe tip against H-3. Look at the display and check the current. I observed 2.77 milliamperes (mA) flowing through the load resistor.

Use Ohm's law to calculate the theoretical voltage E_{load} that should appear across the resistance. Multiply the current I_{load} in milliamperes (thousandth of an ampere) by the load resistance R_{load} in kilohms (thousands of ohms) directly to

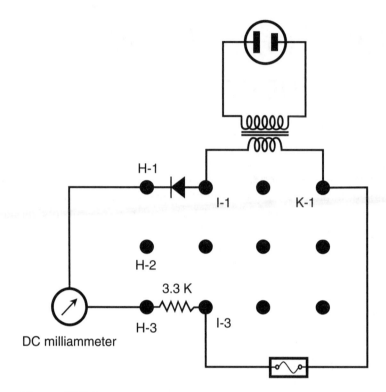

Figure AC11-1   *Measurement of the effective output current through a load from the rectifier alone.*

obtain E_{load} in volts. My arithmetic told me that the theoretical load voltage should have been

$$E_{\text{load}} = I_{\text{load}} R_{\text{load}}$$

$$= 2.77 \times 3.25$$

$$= 9.00 \text{ V}$$

Take the meter probe tips away from H-1 and H-3, and connect a jumper between those terminals. Switch the meter to display DC volts. Measure the actual voltage across the load resistor by placing the meter probe tips between H-3 and I-3. I got 9.04 V, representing a discrepancy of less than 0.5 percent between theory and reality.

Switch the transformer to its higher-voltage output and repeat the above-described tests and calculations. In this situation, my meter indicated a load current of 3.83 mA and a load voltage of 12.44 volts (V), while Ohm's law told me that the theoretical voltage should have been 12.45 V. Again, theory and practice closely agreed.

Battery Aiding Rectifier

With your gloves still on, switch the transformer back to its lower-voltage output, and connect the battery in series with the rectifier output as shown in Fig. AC11-2. The negative battery terminal should go to terminal H-1, and the positive battery terminal should go to H-2, so the battery works with the rectifier. Set the digital meter for DC milliamperes. Hold the negative meter probe tip against terminal H-3, and hold the positive meter probe tip against H-2. Note the meter reading. I got 3.82 mA.

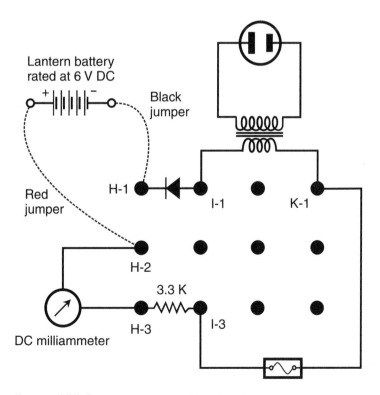

Figure AC11-2 *Measurement of the effective output current through a load from the rectifier and battery in series, with the battery aiding the rectifier.*

Once again, calculate the theoretical voltage E_{load} using Ohm's law. In this situation, my calculation yielded

$$E_{load} = I_{load}R_{load}$$
$$= 3.82 \times 3.25$$
$$= 12.42 \text{ V}$$

Disconnect the meter from H-2 and H-3, and connect a jumper between those terminals. Switch the meter to display DC volts. Measure the actual load voltage by placing the meter probe tips against terminals H-3 and I-3, so the meter is directly across the resistor. I got a reading of 12.43 V. After witnessing unpredictable results in previous tests using a battery in series with a rectifier, I was surprised to see such close agreement between theory and reality.

Switch the transformer to its higher-voltage output and repeat the measurements and calculations. My meter showed a load current of 4.84 mA. I calculated that the load voltage should have been 15.73 V. My meter displayed 15.76 V.

Battery Bucking Rectifier

Keep your gloves on! Once again, switch the transformer to its lower-voltage output. Wire up the battery so that it works against the rectifier as shown in Fig. AC11-3. Set your meter for DC milliamperes. Hold the meter probe tips against terminals H-2 and H-3 to get a positive current reading. When I did this test, my meter displayed 1.88 mA.

Ohm's law predicted that I should have observed a theoretical voltage across the load resistance of

$$E_{load} = I_{load}R_{load}$$
$$= 1.88 \times 3.25$$
$$= 6.11 \text{ V}$$

Disconnect the meter from H-2 and H-3, and replace the meter with a jumper. Then set the meter for DC volts and measure the load voltage by placing the meter directly across the resistor. I got 6.10 V.

Switch the transformer to its higher-voltage output. Do the current and voltage tests and calculations again. My meter told me that the load carried 2.89 mA, Ohm's law predicted that I should have seen a load voltage of 9.39 V, and I measured the actual load voltage as 9.40 V.

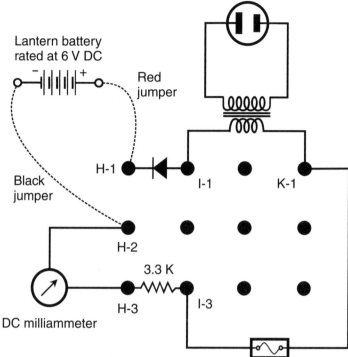

Figure AC11-3 *Measurement of the effective output current through a load from the rectifier and battery in series, with the battery bucking the rectifier.*

Summary of Results

Tables AC11-1 and AC11-2 show the results of my current and voltage measurements, along with my theoretical calculations of the load voltage, in all six scenarios for this experiment. I never saw a discrepancy of more than 0.5 percent between the theoretical and actual load voltages.

Table AC11-1 Measured currents, calculated voltages, and actual voltages I obtained with various combinations of a half-wave rectifier and a 6.44-V lantern battery in series with a 3.25-K load. The transformer output switch was set at the lower-voltage position. For Ohm's law calculations, milliamperes times kilohms equals volts.

Condition	Measured Milliamperes through Load	Calculated Volts across Load	Measured across Load
No battery	2.77	9.00	9.04
Battery aiding	3.82	12.42	12.43
Battery bucking	1.88	6.11	6.10

Table AC11-2   Measured currents, calculated voltages, and actual voltages I obtained with various combinations of a half-wave rectifier and a 6.44-V lantern battery in series with a 3.25-K load. The transformer output switch was set at the higher-voltage position. For Ohm's law calculations, milliamperes times kilohms equals volts.

Condition	Measured Milliamperes through Load	Calculated Volts across Load	Measured across Load
No battery	3.83	12.45	12.44
Battery aiding	4.84	15.73	15.76
Battery bucking	2.89	9.39	9.40

Now Try This!

Conduct the above-described measurements with a 1.5-K resistor as the load. Then do it with two, three, and four 1.5-K resistors in parallel. Remember to test each resistor combination with your ohmmeter before placing the components in service as working loads. How do your calculated voltages (using Ohm's law) across these loads compare with the values you get when you measure the voltages?

Rectifier/Filter and Battery under Load

In this experiment, you'll add a filter capacitor to the rectifier circuit and then measure the load currents and voltages. These tests will reveal how Ohm's law works when you connect a battery in series with the filtered output of a half-wave rectifier to drive current through a known resistance. You'll need everything from Experiment AC11, along with a 1000-microfarad (μF) electrolytic capacitor rated at 35 WVDC.

Set It Up

Switch off the power strip. Set the transformer output switch to its lower-voltage position. Take the capacitor and a 3-inch (in) length of fine, bare copper wire and interconnect the components as shown in Fig. AC12-1. Wrap the capacitor's positive lead around H-1, which also goes to the rectifier's cathode. Wrap the capacitor's negative lead around I-2. Wrap the length of bare wire around I-2 and I-3 to connect the negative side of the capacitor directly to the "non-transformer" side of the fuse. Place a 3.3-kilohm (K) resistor between H-3 and I-3. Before applying power, use your ohmmeter to measure the actual resistance. Mine tested at 3.25 K.

Rectifier/Filter Alone

Put on your gloves and glasses. Double-check to be certain that the transformer is set for its lower-voltage output. Once again, examine the diode and capacitor polarities to be sure that they're correct:

■ The transformer's "non-fuse" wire should go to the diode's anode.

■ The diode's cathode should go to the capacitor's positive lead.

■ The capacitor's negative lead should go to the fuse holder's "non-transformer" wire.

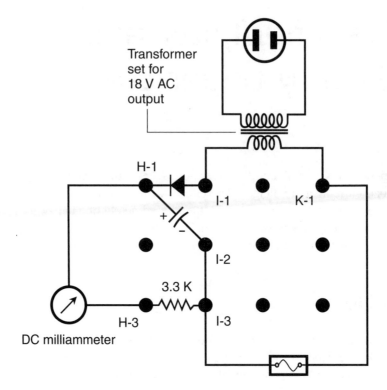

Figure AC12-1   *Measurement of the filtered output current through a load from the power supply alone.*

Plug the transformer into the power strip. Set your meter for DC milliamperes. Apply power to the system. Hold the positive meter probe tip against H-1 and the negative meter probe tip against H-3, as shown in Fig. AC12-1. Check the current. I observed 8.72 milliamperes (mA) through the load resistor.

With Ohm's law, work out the theoretical load voltage E_{load}. When I multiplied the current I_{load} in milliamperes by the resistance R_{load} in kilohms, I obtained

$$
\begin{aligned}
E_{\text{load}} &= I_{\text{load}} R_{\text{load}} \\
&= 8.72 \times 3.25 \\
&= 28.3 \text{ V}
\end{aligned}
$$

Place a jumper between terminals H-1 and H-3. Set your meter for DC volts, and measure the load voltage by holding the probe tips against H-3 and I-3, so the meter is in parallel with the resistor. When I did this test, I got 28.3 volts (V)—a perfect outcome!

Battery Aiding Rectifier/Filter

With your gloves and glasses still on and the transformer still at its lower-voltage setting, connect your lantern battery in series with the rectifier/filter output as shown in Fig. AC12-2, so that the battery and the rectifier/filter aid each other. Set the meter to display DC milliamperes. Place the negative meter probe tip against H-3. Place the positive meter probe tip against H-2. Check the meter reading. My meter said that the load carried 10.66 mA.

Calculate the theoretical voltage E_{load} that should appear across the resistor. My arithmetic yielded

$$E_{\text{load}} = I_{\text{load}}R_{\text{load}}$$
$$= 10.66 \times 3.25$$
$$= 34.6 \text{ V}$$

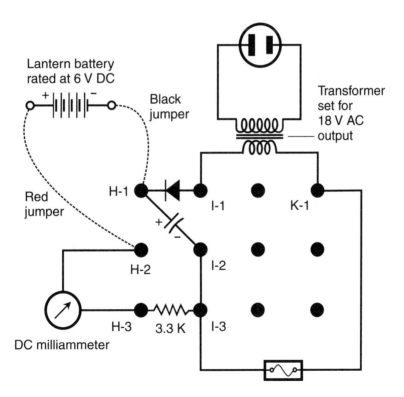

Figure AC12-2   *Measurement of the filtered output current through a load from the power supply and battery in series, with the battery aiding the supply.*

Disconnect the meter from H-2 and H-3, and connect a jumper in the meter's place. Set the meter for DC volts. Measure the load voltage by placing the meter probe tips between H-3 and I-3. I got 34.6 V—perfection again!

Battery Bucking Rectifier/Filter

Keep the transformer output at the low-voltage setting. Leave your gloves and glasses on. Reverse the jumpers from the battery, so that the battery bucks the rectifier/filter (Fig. AC12-3). Set the meter for DC milliamperes, and hold the meter probe tips against H-2 and H-3. My meter said that the resistor carried 6.73 mA.

Once again, use Ohm's law to work out the theoretical voltage across the load resistance. I got

$$E_{load} = I_{load}R_{load}$$
$$= 6.73 \times 3.25$$
$$= 21.9 \text{ V}$$

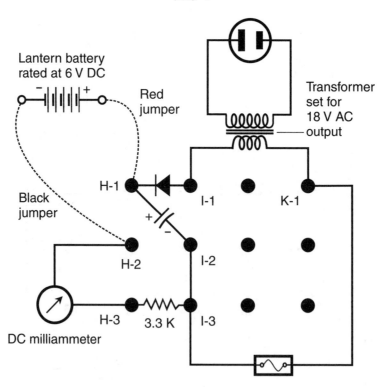

Figure AC12-3 *Measurement of the filtered output current through a load from the power supply and battery in series, with the battery bucking the supply.*

Table AC12-1 Measured currents, calculated voltages, and actual voltages I obtained with various combinations of a half-wave filtered power supply and a 6.44-V lantern battery in series with a 3.25-K load. The transformer output switch was set at the lower-voltage position. For Ohm's law calculations, milliamperes times kilohms equals volts.

Condition	Measured Milliamperes through Load	Calculated Volts across Load	Measured Volts across Load
No battery	8.72	28.3	28.3
Battery aiding	10.66	34.6	34.6
Battery bucking	6.73	21.9	21.9

Disconnect the meter from H-2 and H-3. Place a jumper between those terminals. Set the meter for DC volts, and measure the actual voltage across the load. When I conducted this test, my meter displayed 21.9 V. That made three perfect outcomes in a row!

Summary of Results

Table AC12-1 shows the results of my current and voltage measurements, along with my theoretical calculations of the load voltage, in all three situations for this experiment. No measurable discrepancy existed between the theoretical and real worlds. I hope that your luck is as good as mine was!

Now Try This!

Conduct the above-described calculations and measurements with two, three, and four 3.3-K resistors in parallel as the load. With the circuit powered-down, check each resistor combination with your ohmmeter to be sure that you know the actual load resistances. How do your calculated load voltages (using Ohm's law) compare with the actual load voltages?

AC13

A Full-Wave
Bridge Rectifier

In this experiment, you'll build a rectifier with four diodes in a square matrix called a *bridge*. Along with the breadboard, transformer, and fuse, you'll need four diodes rated at 3 amperes (A) and 400 peak inverse volts (PIV), a resistor rated at 3.3 kilohms (K), some jumpers, your digital meter, and your rubber gloves.

How It Works

In Experiment AC4, you saw how a half-wave rectifier converts AC to pulsating DC by "chopping off" half of the wave cycle. The eliminated part of the input doesn't contribute anything at all to the output; it's completely wasted! A *full-wave bridge rectifier* takes advantage of both halves of the cycle, as shown in Fig. AC13-1. Graph A shows the input AC waveform; graph B shows the rectified output. The entire wave gets through, but the full-wave circuit inverts every other half-cycle. Therefore, both half-cycles produce output of the same polarity.

In theory, given a constant AC input voltage, the RMS pulsating DC output voltage from a full-wave bridge rectifier is twice the RMS pulsating DC output voltage from a half-wave rectifier. That's because the full-wave output has twice as many pulses over time, even though the pulses themselves aren't any more intense. With a full-wave bridge rectifier, the RMS voltages of the input AC and the output DC are theoretically identical. In a practical circuit, because of the small voltage drops across the diodes, the RMS pulsating DC output voltage is slightly lower than the RMS AC input voltage.

Before you start building the rectifier circuit for the following tests, move the transformer output leads to two breadboard terminals that aren't connected to any other components. Put on your gloves, set the transformer to its lower-voltage output, plug it into the power strip, and switch the power strip on. Measure the AC voltage at the transformer output. (You did this in an earlier experiment, but that was probably on a different day, and the household utility AC might not be quite the same now as it was then.) My meter indicated an AC output voltage of 20.8 volts (V) RMS.

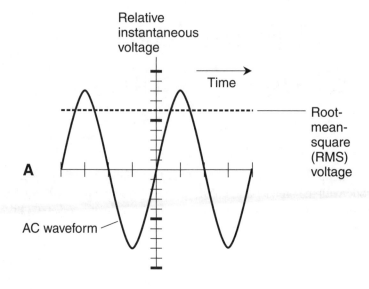

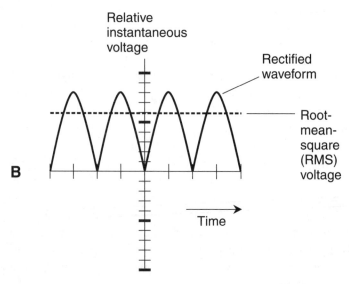

Figure AC13-1   *Full-wave rectification. At A, the AC wave-form as it appears at the power strip. At B, the pulsating DC waveform as it appears at the output of the rectifier. Dashed lines indicate root-mean-square (RMS) or effective voltages.*

Build and Test the Bridge

Switch off the power strip. Arrange the components on the breadboard as shown in Fig. AC13-2. Pay extra-close attention to the polarities of the diodes! Use a 4-inch (in) length of small-gauge bare copper wire to short out terminals I-2, J2, and K2. This "extension" has no electrical significance, but it will give you a little more room on the breadboard than you would otherwise have.

Once you're certain that you have all the components interconnected properly, switch the power strip on. Set the digital meter to indicate DC voltage. Hold the meter probe tips against terminals H-1 and K-2. The positive probe should go to H-1, while the negative probe goes to K-2. Note the meter reading. When I did this

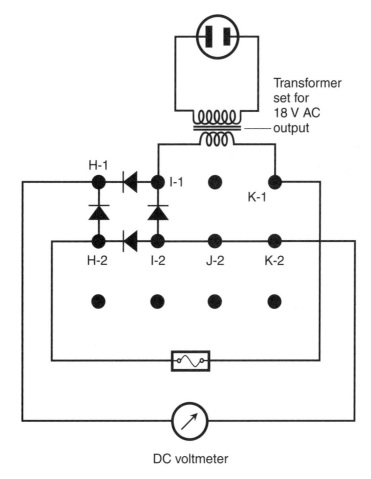

Figure AC13-2 *Measurement of the open-circuit pulsating DC output voltage from the bridge rectifier circuit.*

test, my meter told me that the open-circuit pulsating DC output from the bridge rectifier was 18.6 V RMS, which was 2.2 V less than the RMS AC output voltage from the transformer alone.

For a few moments, I wondered why the voltage difference was so great between the RMS AC input and the RMS pulsating DC output. I expected it to be somewhere between about 0.5 V and 1.2 V, but not anywhere near 2.2 V! Then I looked closely at the component arrangements, and I saw that each half of the wave cycle is forced to go through two diodes in series. Therefore, the voltage gets "bumped down" not once, but twice with every DC pulse. I also realized that the voltage drop across the diodes takes place at *every instant in time* during a *pulsating wave cycle*, so the diodes in this circuit don't behave in quite the same way as they would in a pure DC voltage reducer, such as the one you built in Experiment DC15.

Output Voltage under Load

Switch off the power strip, and then connect a 3.3-K resistor between terminals H-3 and I-3 as shown in Fig. AC13-3. Connect a 3-in length of fine, bare copper wire between terminals I-2 and I-3. In this arrangement, the resistor serves as a load for the rectifier output. Measure the actual value of the resistor before connecting terminal H-3 to any other components. My resistor tested at 3.24 K.

Place a jumper between terminals H-1 and H-3 to connect the load resistor directly across the rectifier output. Switch on the power strip. Set your meter to indicate DC volts. Place the meter's positive probe tip against terminal H-1, and place the negative probe tip against K-2 to get a reading of the RMS pulsating DC output voltage across the load resistor. I got 17.7 V, telling me that the presence of the load dropped the rectifier output by almost a full volt.

Output Current through Load

Use Ohm's law to calculate the theoretical current that the full-wave bridge rectifier should drive through the load resistor. I measured a voltage of $E = 17.7$ V and a resistance of $R = 3.24$ K (which I converted to 3240 ohms). I therefore expected the current I to be

$$I = E/R$$
$$= 17.7/3240$$
$$= 0.00546 \text{ A}$$
$$= 5.46 \text{ mA}$$

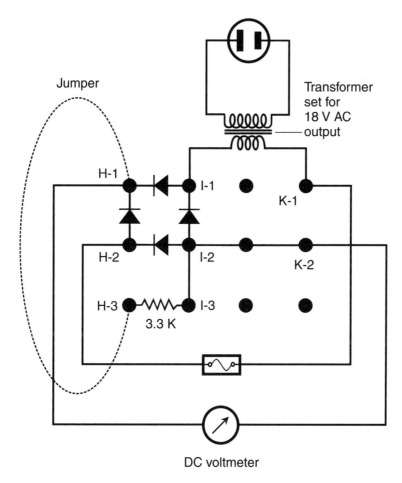

Figure AC13-3 *Measurement of the pulsating voltage from the bridge rectifier circuit across a resistive load.*

Switch off the power strip. Remove the jumper from between terminals H-1 and H-3. Set your digital meter to indicate DC milliamperes (mA). Switch the power strip back on, and then connect the meter where the jumper was, as shown in Fig. AC13-4. The positive meter probe tip should go to H-1, and the negative probe tip should go to H-3. What does your meter say? I was pleased to see 5.46 mA, exactly the same as the theoretical value that I calculated.

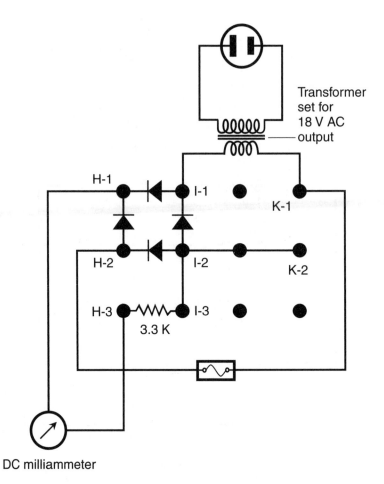

Figure AC13-4   *Measurement of the pulsating current that the bridge rectifier circuit drives through a resistive load.*

Now Try This!

Conduct the above-described calculations and measurements with a 1.5-K resistor as the load. Then do the same maneuvers with two, three, and four 1.5-K resistors in parallel. Using your ohmmeter, measure the value of each resistor combination ahead of time with the circuit powered-down and terminal H-3 connected to no other components. How does the rectifier's output voltage change as the load resistance goes down? How does your calculated load current compare with the actual load current in each situation?

AC14

A Filtered Full-Wave Power Supply

In this experiment, you'll add a 1000-microfarad (μF) electrolytic capacitor, rated at 35 working volts DC (WVDC), to the full-wave bridge rectifier to get rid of the ripple in the output. You'll need all the components you used in Experiment AC13 along with the capacitor, a couple of extra jumpers, your rubber gloves, and your safety glasses.

Full-Wave versus Half-Wave filtering

In Experiment AC8, you placed a filter capacitor across the output of a half-wave rectifier. In that type of rectifier circuit, the capacitor dramatically increases the effective output voltage with an open circuit or a high load resistance. The voltage across the capacitor rises to, and remains near, the peak rectifier output voltage, not the RMS voltage. (Look back at Fig. AC8-2 on page 191 to see a graphical illustration of the effect.) In a half-wave rectifier circuit, the peak output voltage is far greater than the RMS output voltage. That's why the capacitance increases the voltage so much.

The same thing will happen in this experiment when you place a filter capacitor across the output of a full-wave bridge rectifier, but not to such a great extent, because the RMS output voltage of a full-wave bridge rectifier is twice that of a half-wave rectifier receiving the same AC input. There's another difference here, as well. The capacitor receives charging pulses twice as often with full-wave rectification (Fig. AC14-1) as compared with half-wave rectification (Fig. AC8-2). As a result, the output of the full-wave rectifier is easier to filter.

Despite the difference in the RMS output voltages, and despite the fact that the capacitor gets charged up twice as often, the *peak* output voltage from a full-wave

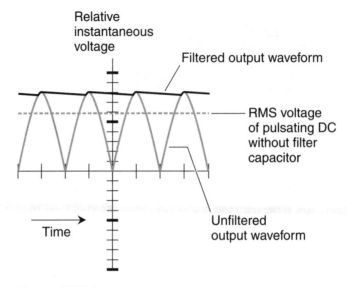

Figure AC14-1   *A filter capacitor "smooths" the output of a full-wave rectifier by charging with each pulse, holding the voltage near the peak value.*

bridge rectifier is the same as the peak output voltage from a half-wave rectifier getting the same AC input. Therefore, it's reasonable to expect that the capacitor should charge up to about the same voltage here as it did in the half-wave situation.

Install the Capacitor

Begin with the breadboard as you left it at the end of Experiment AC13. Keep the transformer switch set to the lower-voltage position. Put on your rubber gloves and safety glasses. Switch the power strip on. Set your digital meter set to measure DC voltage, and use it to verify the rectifier output polarity at terminals H-1 and I-2. Terminal H-1 should be positive, and terminal I-2 should be negative. Once you've checked the voltage, switch the power strip off.

Install the electrolytic capacitor with the negative lead at terminal J-2 and the positive lead at J-3. As in Experiment AC8, be certain that you get the capacitor's polarity correct. Connect a jumper between terminals J-3 and H-1 to place the capacitor directly across the output of the full-wave bridge rectifier. The capacitor's positive lead should go to the cathodes of two diodes (the ends with the stripes); the capacitor's negative lead should go to the anodes of the other two diodes as shown in Fig. AC14-2.

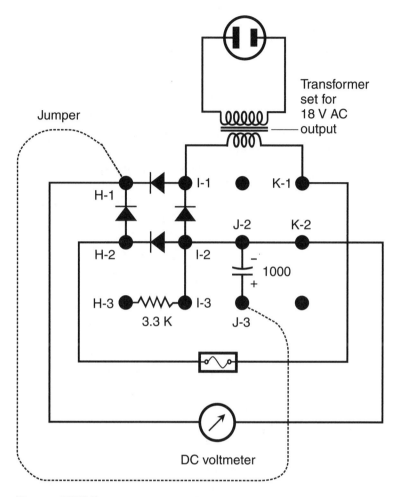

Figure AC14-2 *Measurement of the open-circuit DC output voltage from the filtered full-wave bridge power supply.*

Open-Circuit Voltage

Switch on the power strip. Set your digital meter to display a moderate to high DC voltage. Place the meter's negative probe tip against terminal K-2, and place the positive probe tip against H-1, so that you can read the open-circuit filtered output voltage of the rectifier-capacitor combination. When I conducted this test, I got 28.6 volts (V), almost exactly the same as the filtered voltage in the half-wave situation of Experiment AC8. This result was what I expected, as explained above.

Voltage under Load

Switch off the power strip. Use a jumper to discharge the capacitor. Remove the jumper from terminal H-1 and connect that end to H-3, so it shorts J-3 (the positive side of the capacitor) to H-3 (the "non-diode" side of the load resistor). Place another jumper between H-3 and H-1 to connect the load resistor directly across the output of the rectifier-filter circuit.

Switch on the power strip. The complete power supply is now working into a resistive load. With your meter still set to read DC volts, measure the voltage across the load as shown in Fig. AC14-3. My meter displayed 27.7 V DC with the

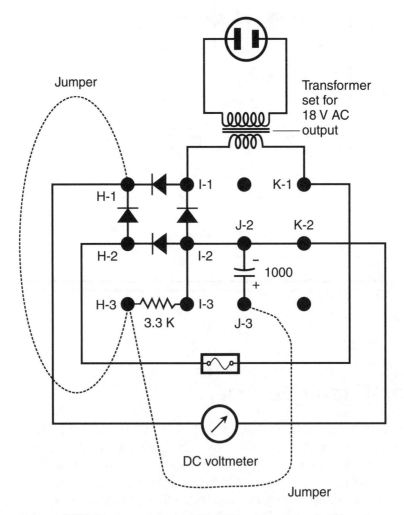

Figure AC14-3 *Measurement of the DC voltage from the filtered full-wave bridge power supply across a resistive load.*

3.24-kilohm (K) load, a decrease of 0.9 V from the open-circuit situation—but identical to the filtered output of the half-wave rectifier working into the same load during Experiment AC8.

Current through Load

Once again, switch the power strip off. Discharge the capacitor by shorting it out with a jumper. Set your meter to display DC milliamperes (mA). Remove both jumpers from the breadboard completely. Then put a new jumper between terminals H-1 and J-3. Switch the power strip back on. Hold the negative meter probe tip against H-3, and hold the positive probe tip against J-3 to measure the current through the load resistor as shown in Fig. AC14-4. When I did this test, I got 8.50 mA.

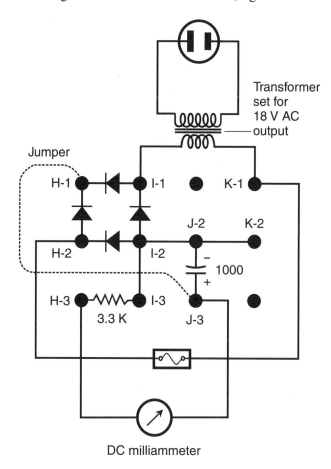

Figure AC14-4   *Measurement of the current that the filtered full-wave bridge power supply drives through a resistive load.*

Now, just as you did in Experiment AC8, check the voltage, current, and resistance for adherence to Ohm's law with the formula

$$I = E/R$$

where I represents the current in amperes, E represents the voltage in volts, and R represents the resistance in ohms. My measured values, converted to standard units, were precisely in agreement with those of Experiment AC8:

$$E_{load} = 27.7 \text{ V}$$

$$R_{load} = 3240 \text{ ohms}$$

$$I_{load} = 0.00850 \text{ A}$$

According to the formula, I should have seen a current through the resistor of

$$I_{load} = E_{load}/R_{load}$$

$$= 27.7/3240$$

$$= 0.00855 \text{ A}$$

Now Try This!

Conduct the above described voltage measurement with two, three, and finally four 3.3-K resistors in parallel as the load. How do these results compare with those you got when you did the same tests at the end of Experiment AC8?

AC15

How Bleeders Work

Large-value capacitors work well as power-supply filters because they can hold a lot of electrical charge for a long time. But this phenomenon isn't entirely good. Unless there's a load connected to the supply output, a filter capacitor remains charged after power-down. In a high-voltage supply, this *residual charge* poses a shock hazard. To ensure that the capacitor doesn't stay charged indefinitely, a *bleeder* should be connected across it. In this experiment, you'll see some bleeders in action.

Hold That Charge!

Put on your gloves and glasses. Switch off the power strip. Ensure that the capacitor is fully discharged by shorting it out with a jumper for a couple of seconds. Arrange the components on your breadboard as shown in Fig. AC15-1. As always, pay attention to the capacitor polarity. Set your meter to indicate DC volts. Insert the positive meter probe tip vertically into the "alligator clip" at terminal H-1, so that the probe stays in place without your having to hold it there. In the same fashion, insert the negative meter probe tip into the clip at terminal I-3.

Be sure that the transformer switch is in its lower-voltage position. Switch on the power strip for a few seconds, and then switch it off again. Watch the meter display. While power is applied, the meter reading should be approximately the same as it was when you tested the filter capacitor under no-load conditions in Experiment AC14. On this occasion, I observed 28.5 volts (V). After you remove power, the meter reading should gradually decline—but you'll know before long that it will be quite awhile before it gets down to zero. Are you surprised to find out how well the capacitor holds its charge after power has been removed?

Choosing a Resistance

If you connect a resistor across the filter capacitor, you can force the capacitor to discharge in a reasonable time after power-down without placing an excessive load on the power supply while it's in use. You must be sure that the resistance is large

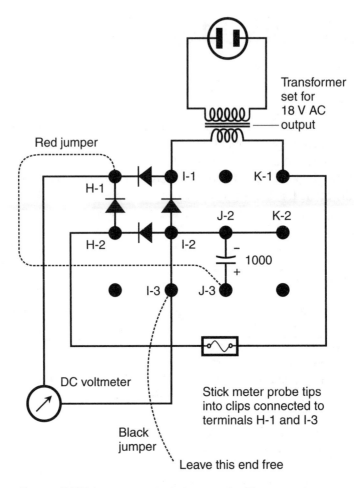

Figure AC15-1   *Testing the behavior of a filter capacitor after removal of power under no-load conditions.*

enough so that the resistor won't burn out, but not so large that the capacitor takes too long to lose its charge.

Assuming you have the Radio Shack resistors that I've recommended, your resistors are all rated to dissipate up to 0.5 watt (W) or 500 milliwatts (mW) of power continuously and indefinitely. Knowing this fact, we can calculate the minimum safe resistance for our bleeder. From basic electricity theory, we recall that

$$P = E^2/R$$

where P is the power in watts, E is the voltage in volts, and R is the resistance in ohms. We can rearrange this formula to get

$$R = E^2/P$$

The capacitor holds a sustained voltage of a little less than 30 V when power is applied to this circuit. Let's say that $E = 30$ V to give us a little margin of safety. We know that $P = 0.5$ W. Therefore

$$R = 30^2/0.5$$
$$= 900/0.5$$
$$= 1800 \text{ ohms}$$

The smallest bleeder resistance that we can use without overheating the resistor is 1800 ohms or 1.8 kilohms (K). If you bought all of the resistors listed in Table DC1-1 (but no more), then 3.3 K is the smallest resistor value in your collection that "qualifies."

You should have five 3.3 K, $1/2$-W resistors on hand. You can connect any number of these identical resistors in series, or any number of them in parallel, and the combination will stay within the safety margin for power dissipation in this particular power-supply circuit. (Can you mathematically show why this is true?)

Bleed off the Charge

Be sure that the power strip is switched off. Discharge the filter capacitor by shorting it out with a jumper for a couple of seconds. Then wire up five 3.3-K resistors in series among the six terminals F-6 through K-6, as shown in Fig. AC15-2. Connect a red jumper between J-3 and K-6, but don't install the black jumper between I-3 and I-6 yet. Use your ohmmeter to measure the individual values of the resistors to be sure that none of them are "errant units."

Once you've determined that all the resistors are within their rated values, place your ohmmeter terminals between terminals I-6 and K-6 to measure the value of the two-resistor series combination between those terminals. When I did this test, I got 6.49 K. Connect the black jumper between terminals I-3 and I-6 to place the two-resistor series combination across the filter capacitor.

Set your meter to indicate DC volts. Insert the meter probe tips into the clips at terminals H-1 and I-3 so that they stay in place by themselves. Switch on the power strip. Note the voltage. It will probably be a little lower than it was when there were no resistors in parallel with the capacitor. I got 27.9 V.

The Discharge Decrement

Find a clock or wristwatch with a moving second hand (or a digital seconds display) that's easy to read. Switch off the power strip. Check and write down the voltage after 1 second (s) passes. Switch the power strip back on for a moment.

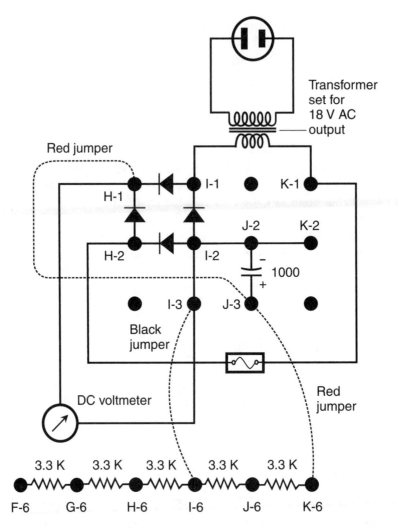

Figure AC15-2 *Testing the behavior of a filter capacitor after removal of power with a bleeder resistance to speed up the discharge process.*

Then switch it off again; check and write down the voltage after 2 s has gone by. Repeat this process as many times as necessary to get voltage readings for elapsed times of 3, 4, 5, 10, 15, 20, 25, 30, 40, and 50 s. Compile the data and arrange it in tabular form. Table AC15-1 shows my results.

Now plot the data from your voltage-vs.-time table in the form of a graph. The horizontal axis should show the time (in seconds) after power-down. The vertical axis should show the voltage across the capacitor-bleeder combination. Figure AC15-3 shows my graph, based on the data from Table AC15-1. The small open circles represent the actual data points. The smooth curve approximates the *discharge decrement* for the capacitor-resistor combination.

Table AC15-1   Measured voltages across a 6.49-K bleeder resistor in parallel with a 1000-microfarad (µF) capacitor, as a function of the time in seconds after removal of power. The transformer output switch was set at the lower-voltage position.

Seconds after Power-Down	Volts across Bleeder Resistor
0	27.9
1	24.3
2	19.5
3	17.0
4	14.7
5	12.8
10	6.1
15	3.2
20	1.7
25	1.2
30	0.7
40	0.4
50	0.2

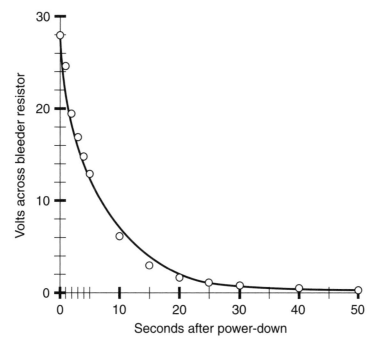

Figure AC15-3   *Graph of the data from Table AC15-1, showing how the filter capacitor discharges through a bleeder resistance.*

Now Try This!

Move the black jumper from terminal I-6 to terminal J-6, so that only one 3.3-K resistor is connected across the filter capacitor. Perform all of the previous tests and measurements again. Then move the black jumper in turn to terminals H-6, G-6, and F-6, so that your bleeder consists of three, four, and five 3.3-K resistors in series. Compile tables and discharge graphs for each situation. How do they compare?

A Zener-Diode
Voltage Regulator

The DC output voltage from a filtered rectifier varies in proportion to the AC input voltage unless *voltage regulation* is employed. In this experiment, you'll see how a device called a *Zener diode* can regulate the output of a power supply by working as a "brute-force" *voltage limiter*. You'll need everything from the previous few experiments along with a Zener diode rated at 5.1 volts (V), and another Zener diode rated at 12 V.

What's a Zener Diode?

Have you learned that a diode can never conduct any current when it's reverse-biased (the cathode is positive with respect to the anode)? That's an oversimplification! If the reverse voltage across a diode's P-N junction gets high enough, the voltage overcomes the ability of the junction to block the flow of current. Then the junction conducts, a phenomenon called *avalanche effect*. If the reverse-bias voltage drops back below the critical value or so-called *avalanche voltage*, the junction blocks the current again.

The avalanche voltage of an ordinary semiconductor diode is much higher than the reverse bias voltage ever gets. But a Zener diode's avalanche voltage can be quite low. For example, suppose that a certain Zener diode has an avalanche voltage, also called the *Zener voltage*, of 12 V. If reverse bias is applied, the diode acts as an open circuit as long as the voltage is less than 12 V. But as soon as it reaches 12 V, the diode conducts, preventing the potential difference across the device from exceeding 12 V.

Set Up the Circuit

Figure AC16-1 illustrates the breadboard layout for constructing a full-wave, filtered power supply with a simple Zener-diode voltage regulation circuit. The schematic symbol for a Zener diode (D in the diagram) looks like the symbol for an ordinary diode, except that the cathode line is "bent" at the ends.

Be sure you're wearing your gloves and glasses. Switch off the power strip. Use a jumper to briefly discharge the filter capacitor. Find a 5.1-V Zener diode. Note the diode polarity! The anode, represented by the end *without* the black stripe, should be connected to the negative rectifier output. You might need a magnifying glass to see the stripe, which represents the cathode. Connect the cathode to terminal K-3. Wrap the leads around the nails carefully so you don't break the diode's fragile glass case.

Remove the shorting wire from between terminals I-2 and I-3. Install a 1.5-kilohm (K) resistor (R in the figure) between terminals I-3 and J-3. Use the ohmmeter to test its value. My resistor checked out at 1.47 K. This resistor limits the current that can flow through the Zener diode. Without the limiting resistor, the diode would burn out.

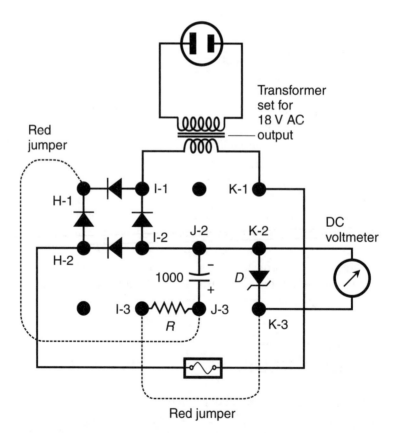

Figure AC16-1 *Addition of a current-limiting resistance R and a Zener diode D to regulate the DC output voltage from a filtered full-wave rectifier. The meter is connected to measure the regulated output voltage under no-load conditions.*

Output with 5.1-V Zener

Connect a red jumper between terminals H-1 and J-3, and then connect another red jumper between I-3 and K-3 as shown in Fig. AC16-1. Set your meter to read DC volts. Switch on the power strip. Hold the negative meter probe tip against terminal K-2, and hold the positive probe tip against K-3. Note the voltage across the Zener diode. It should be close to 5.1 V. When I did this test, I got 5.15 V.

Take the voltmeter probe tips away from K-2 and K-3, and remove the red jumper from between I-3 and K-3. Set the meter for DC milliamperes (mA). Hold the negative meter probe tip against terminal K-3, and hold the positive probe tip against I-3 as shown in Fig. AC16-2. The meter should indicate the current through the Zener diode D and the limiting resistor R. When I conducted this test, I got 15.03 mA.

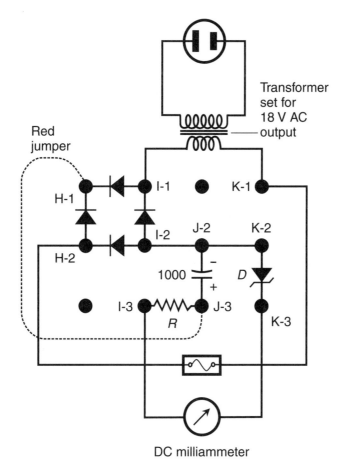

Figure AC16-2 *Measurement of the current through the diode and the limiting resistor under no-load conditions.*

Output with 12-V Zener

Take the meter away from terminals I-3 and K-3. Power-down the system by switching the power strip off. Discharge the filter capacitor with a jumper. Replace the 5.1-V Zener diode with a 12-V Zener diode. Connect the red jumper back between terminals I-3 and K-3, and set your meter to DC volts again. Switch on the power strip and place the meter probe tips against K-2 and K-3, so you have the arrangement of Fig. AC16-1 once more. The meter should display about 12 V. Mine showed 11.69 V.

Remove the meter from K-2 and K-3. Disconnect the jumper from I-3 and K-3. Set the meter for DC milliamperes. Hold the meter probe tips against K-3 and I-3 to measure the current through D and R (Fig. AC16-2) again. My meter said that the current was 10.66 mA.

Now Try This!

Whenever you use a Zener diode in conjunction with a limiting resistor, you must not exceed the resistor's power-dissipation rating. If you select a resistance that's too low, you'll burn out the resistor. You'll also risk exceeding the diode's current rating, destroying it as well!

You can use the standard DC formula for power in terms of current and resistance to calculate the wattage dissipated by R in the circuits you've just finished testing. That formula is

$$P = I^2 R$$

where P is the power in watts, I is the current in amperes, and R is the resistance in ohms. Remember to use your measured value of R, not the resistor's rated value. Also note that in this formula, you *must* convert I to amperes and R to ohms in order to calculate the power in watts.

A Zener-Diode
Voltage Reducer

In this experiment, you'll find out what happens when you connect a 5.1-volt (V) Zener diode in series with the output of a power supply that already employs a 12-V Zener diode in parallel to regulate the voltage. You'll need all the components from Experiment AC16, along with a 3.3-kilohm (K) resistor and an extra jumper.

One Diode without Load

Switch off the power strip. Put on your gloves and glasses. Discharge the filter capacitor with a jumper. Arrange the components on your breadboard as shown in Fig. AC17-1. This diagram represents the same circuit arrangement as Fig. AC16-1, except that now the current-limiting resistor and Zener-diode specifications are labeled.

Once you're sure that everything is connected properly and all the polarities are right, switch on the power strip. Set the meter to read DC volts. Hold the meter probe tips against terminals K-2 and K-3 to measure the output voltage. Your reading should be just about the same as it was in Experiment AC16 with the 12-V Zener diode. My meter indicated 11.71 V.

Two Diodes without Load

Think all the way back to Experiment DC15. In that set of tests, you saw how an *ordinary diode*, when series-connected in the forward direction, reduces the DC voltage from a battery. What do you think will happen if you connect a *Zener diode* in series, but in the *reverse* direction? I suspected that it would reduce the output of our filtered, regulated power supply by an amount equal to the rated Zener voltage. That seems reasonable, doesn't it? Well, things aren't quite that simple.

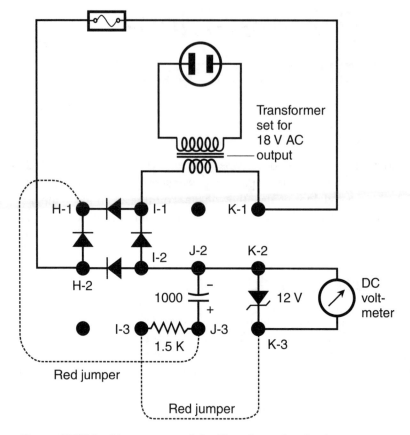

Figure AC17-1 *Measurement of the filtered, regulated voltage across a 12-V Zener diode under no-load conditions.*

Switch off the power strip and discharge the filter capacitor. Place a 5.1-V Zener diode between terminals H-3 and I-3 as shown in Fig. AC17-2. Note the diode's polarity; its cathode (the end with the dark band) should go to the "non-capacitor" end of the limiting resistor. With the meter set for DC volts, switch on the power strip and place the meter probe tips against H-3 and K-2. I expected that in this case, the voltage would be 5.1 V less than it was in the arrangement of Fig. AC17-1, or approximately 6.61 V. But I measured 9.35 V, representing a voltage drop of only 2.36 V.

Evidently, when a Zener diode is connected in series with the output of a regulated, filtered power supply, and when that diode doesn't carry any current (because there's no load), the diode doesn't necessarily behave according to its specifications.

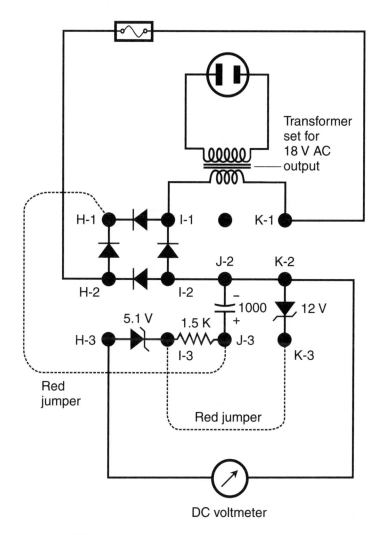

Figure AC17-2   *Measurement of the filtered, regulated, reduced voltage with a 12-V Zener diode in parallel and a 5.1-V Zener diode in series under no-load conditions.*

One Diode with Load

Switch off the power strip and discharge the filter capacitor again. Take a 3.3-K resistor from your collection, and test it with your ohmmeter to be sure its value is reasonably close to 3.3 K. Then place the resistor between terminals K-2 and K-3 so it's in parallel with the 12-V Zener diode as shown in Fig. AC17-3. Switch your meter to read DC volts.

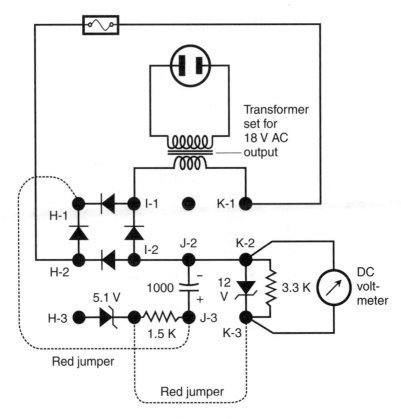

Figure AC17-3 *Measurement of the voltage across a 3.3-K load with a 12-V Zener diode in parallel.*

Caution! *Always double-check to be sure that your meter is set to measure the correct parameter before you connect its probe tips to any of the circuit terminals. When I did the foregoing test for the first time, I accidentally left my meter set to measure resistance, not voltage. The instant that I put the meter into the circuit, the meter's internal fuse blew. I had to take the meter apart, find the fuse, remove it, and go to the local hardware store to buy a new fuse. The sales person had a difficult time tracking down the right type of fuse. I was lucky to get out of there without having to buy a whole new meter.*

Once you're certain that the meter is set to measure DC voltage, hold its probe tips against terminals K-2 and K-3 to measure the regulated, filtered power supply's output voltage with the 3.3-K load. When I performed this test, my meter indicated 11.63 V. The presence of the load resistor dropped the output by 0.08 V compared with the voltage under no-load conditions.

Two Diodes with Load

Once again, switch off the power strip and discharge the filter capacitor with a jumper. Remove the resistor from across the 12-V Zener diode. Connect the resistor between terminals G-3 and H-3 as shown in Fig. AC17-4. The 5.1-V Zener diode should still be between H-3 and I-3. Keep the meter set for DC volts. Switch on the power strip and place the meter probe tips against G-3 and H-3 to measure the voltage across the load. I got 6.82 V, representing a decrease of 4.81 V from the voltage with the 12-V Zener diode alone under load, and a drop of 4.89 V compared with the situation with the 12-V Zener diode alone without a load. The addition of

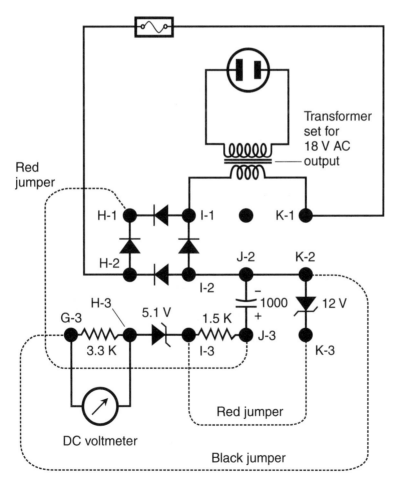

Figure AC17-4 *Measurement of the voltage across a 3.3-K load with a 12-V Zener diode in parallel and a 5.1-V Zener diode in series.*

the load caused the series-connected Zener diode to behave in much better accordance with its rating, but it still didn't quite attain its full specification of 5.1 V.

Now Try This!

Repeat the above experiments with the 5.1-V Zener diode connected the other way, so that it's in the forward direction rather than in the reverse direction. How much does the voltage go down under no-load conditions in this case? What about the voltage with a 3.3-K load?

An AC Spectrum
Monitor

This experiment is easy to perform, but the theory behind it is sophisticated. You'll need a personal computer with an Internet connection. You'll also need a 12-foot (ft), two-wire audio cord with a $1/8$-inch (in) monaural phone plug on one end and spade terminals on the other end (Radio Shack part number 42-2454). If you can't find that component, you can connect a 12-ft length of two-wire speaker cable to a $1/8$-in mono phone plug.

Electromagnetic Fields

The current flowing through the utility grid produces obvious effects on appliances: glowing lamps, blowing fans, and chattering television sets. This AC also produces *electromagnetic* (EM) *fields* that aren't apparent to the casual observer. The presence of this EM energy causes tiny currents to flow or circulate in any object that conducts electricity, such as a wire, a metal rain gutter, the metal handle of your lawn mower, and your body. You're about to deploy a device that can detect this energy and produce a multidimensional graphic display of its characteristics.

Any current-carrying wire is surrounded by theoretical *electric flux lines* and *magnetic flux lines*. Around a straight span of wire, the electric or E flux lines run parallel to the wire and the magnetic or M flux lines surround the wire (Fig. AC18-1A). If the wire carries constant DC, the electric and magnetic fields are *static*, meaning that the E and M fields stay the same all the time. If the wire carries AC or pulsating DC, the fields fluctuate. The varying E field gives rise to a changing M field, which in turn generates another varying E field. As the E and M fields regenerate each other, a "hybrid field" (the EM field) travels away from the wire, perpendicular to both sets of flux lines at every point in space as shown in Fig. AC18-1B.

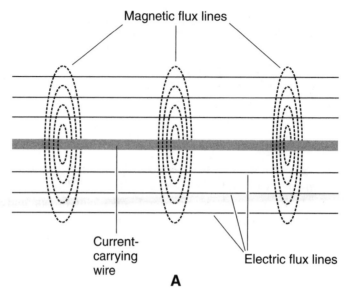

Magnetic flux lines

Current-
carrying
wire

Electric flux lines

A

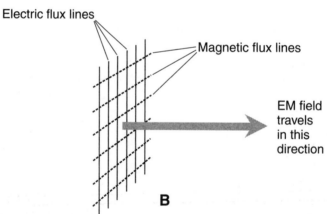

Electric flux lines

Magnetic flux lines

EM field
travels
in this
direction

B

Figure AC18-1   *At A, the electric and magnetic lines of flux around a straight, current-carrying wire. At B, the flux lines far from a current-carrying wire. The EM field travels straight away from the wire, in a direction perpendicular to both sets of flux lines.*

All EM fields display three independent properties: *amplitude, wavelength,* and *frequency*. The amplitude is the intensity of the field. The frequency is the number of full cycles per second. The wavelength is the distance in space between identical points on adjacent wavefronts. At 60 hertz (Hz), the AC utility frequency in the United States, EM waves are 5000 kilometers (km) [approximately 3100 miles (mi)] long in *free space* (air or a vacuum).

The Software

You can use your computer to "look" at the EM fields permeating the space all around you. A simple computer freeware program called *DigiPan*, available on the Internet, can provide a real-time, moving graphical display of EM field components at frequencies ranging from 0 Hz (that is, DC) up to 5500 Hz. Here's the Web site: www.digipan.net.

DigiPan shows the frequency along the *x* axis (horizontally), while time is portrayed as downward movement along the *y* axis (vertically). Figure AC18-2 illustrates this display scheme. The relative intensity at each frequency appears as a color. If there's no energy at a particular frequency, there's no line at all, and the display is black. If there's a little bit of energy at a particular frequency, you'll see a thin, vertical blue line creeping straight downward along the *y* axis. If there's a moderate amount of energy, the line turns yellow. If there's a lot of energy, the line becomes orange or red. The entire display is called a *waterfall*.

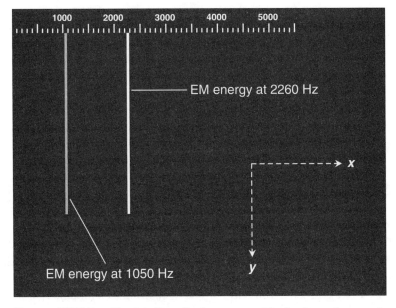

Figure AC18-2   *The DigiPan display system. The horizontal or x axis portrays frequency. The vertical or y axis portrays time. The EM-field "signals" show up as steadily lengthening vertical lines. This drawing shows two hypothetical examples.*

DigiPan is intended for digital communication in a mode called *phase-shift key-ing* (PSK). This mode is popular among amateur radio operators. You can read more about this interesting form of communications by "googling" it. Luckily for us, DigiPan can function as a very-low-frequency *spectrum monitor*, showing the presence of AC-induced EM fields not only at 60 Hz (which you should expect) but at many other frequencies (which you might not expect until you see the evidence). DigiPan doesn't take much computer processing power. Nearly all laptops or desktops can run it handily. If you have a good Internet connection, DigiPan will download and install in a minute or two.

The Hardware

To observe the EM energy on your computer, you'll need an antenna. Cut off the U-shaped spade lugs from the audio cord with a scissors or diagonal cutter. Separate the wires by pulling them apart along the entire length of the cord, so that you get a $1/8$-in monaural phone plug with two 12-ft wires attached.

Insert the phone plug into the *microphone* input of your computer. Arrange the two 12-ft wires so that they run in different directions from the phone plug. You can let the wires lie anywhere, as long as you don't trip over them! This arrangement will make the audio cord behave as a *dipole antenna* to pick up EM energy.

Open the audio control program on your computer. If you see a microphone input volume or sensitivity control, set it to maximum. If your audio program has a "noise reduction" feature, turn it off. Then launch DigiPan and

- Click on "Options" in the menu bar and uncheck everything except "Rx."

- Click on "Mode" in the menu bar and select "BPSK31."

- Click on "View" in the menu bar and uncheck everything.

- Click on "Configure" in the menu bar, select "Waterfall drive," select "Microphone," set the sliding balance control at the center, and maximize the volume.

Once you've carried out these steps, the upper part of your computer display should show a chaotic jumble of text characters on a white background. The lower part of the screen should be black with a graduated scale at the top, showing numerals 1000, 2000, 3000, and so on. Using your mouse, place the pointer on the upper border of the black region and drag that border upward until the white region with the distracting text vanishes.

If things work correctly, you should have a real-time panoramic display of EM energy from zero to several thousand hertz. Unless you're in a remote location far away from the utility grid, you should see a set of gradually lengthening, vertical lines of various colors. These lines represent EM energy components at specific frequencies. You can read the frequencies from the graduated scale at the top of the screen. Do you notice a pattern?

Fundamental and Harmonics

A pure AC sine wave appears as a single *pip* or vertical line on the display of a spectrum monitor (Fig. AC18-3A). This means that all of the energy in the wave is concentrated at one frequency, known as the *fundamental frequency*. But many, if not most, AC utility waves contain *harmonic* energy along with the energy at the fundamental frequency.

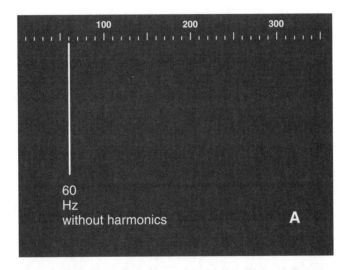

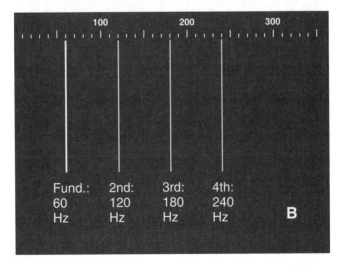

Figure AC18-3   *At A, a spectral diagram of pure, 60-Hz EM energy. At B, a spectral diagram of 60-Hz energy with significant components at the second, third, and fourth harmonic frequencies.*

A harmonic frequency is a whole-number multiple of the fundamental frequency. For example, if 60 Hz is the fundamental frequency, then harmonics can exist at 120 Hz, 180 Hz, 240 Hz, and so on. The 120-Hz wave is at the *second harmonic*, the 180 Hz wave is at the *third harmonic*, the 240-Hz wave at is the *fourth harmonic*, and so on. In general, if a wave has a frequency equal to n times the fundamental where n is some whole number, then that wave is called the *nth harmonic*. (The fundamental is identical to the first harmonic.) In Fig AC18-3B, a wave is shown along with its second, third, and fourth harmonics, as the entire "signal" would appear on a spectrum monitor.

When you look at the EM spectrum display from zero to several thousand hertz using the arrangement described here, you'll see that utility AC energy contains not only the 60-Hz fundamental, but *many* harmonics. When I saw how much energy exists at the harmonic frequencies in my house, I was amazed.

Now Try This!

Place a vacuum cleaner near your EM pickup antenna. Switch the appliance on while watching the DigiPan waterfall. When the motor first starts up, do curves suddenly appear, veering to the right and then straightening out as vertical lines? Those contours indicate energy components that increase in frequency as the motor "revs up" to its operating speed and maintain constant frequencies thereafter. When the motor loses power, do the motor's vertical lines curve back toward the left before they vanish? Those curves indicate falling frequencies as the motor slows down. Try the same tests with a hair dryer, an electric can opener, or any other appliance with an electric motor. Which types of appliances are the "noisiest"? Which are "quietest"?

Part 3
Magnetism

MAG1

Your Magnetism Lab

Keep your breadboard and all the components from the DC and AC experiments. Table MAG1-1 lists the new hardware that you'll need to conduct the experiments in this section. As always, your local Radio Shack retail store, the Radio Shack Web site, or a common hardware store should stock what you seek. If you can't find a particular item in your locale, you can get it (or its equivalent) from one of the mail-order outlets listed in the back of this book.

> **Warning!** *Never touch any open connection or component while it carries, either directly or indirectly, household utility power. The voltage from a "wall outlet" can drive a deadly electrical current through your body.*

> **Caution!** *Use rubber gloves for wire-wrapping or splicing operations if the voltage at any exposed point might exceed 10 volts.*

> **Caution!** *Wear safety glasses at all times as you conduct these experiments, even if you think you don't need them.*

Table MAG1-1   Components list for the magnetism experiments. You can find these items at retail stores near most locations in the United States. Abbreviations: in = inches, ft = feet, AWG = American Wire Gauge, V = volts, and W = watts.

Quantity	Store Type or Radio Shack Part Number	Description
1	See Part 1	All components from Part 1
1	See Part 2	All components from Part 2
1	64-1888	Package of five ceramic disk magnets, 1.125-in diameter
1	Department store or your kitchen	Common metal silverware set (spoon, knife, fork)
1	Department store or your kitchen	Cast-iron skillet
1	Grocery store or your pantry	Various cans of seafood such as tuna, shrimp, sardines, and clams
1	Grocery store	Small roll of aluminum foil
1	Grocery store	Paper plate
1	Grocery store	Plastic plate
1	Grocery store or your kitchen	Opener for large cans (such as tuna)
1	Department store or your home	Small flashlight with fresh batteries
1	Hardware store	Pair of safety glasses (in addition to the pair you bought for DC experiments)
2	Your home	Folding chairs or lawn chairs
1	Hardware store	4-ft length of AWG No. 8 soft-drawn copper wire
2	Hardware store	Threaded steel rods, $1/2$-in diameter, 12 in long
4	Hardware store	Nuts to fit threaded rods
1	Hardware store	20-ft length of 2-wire lamp cord with AWG No. 16 conductors
1	Hardware or department store	Electric heater with two or three settings, maximum 1500 W at 117 V AC
1	Hardware or home supply store	Single-speed, high-velocity fan for use with 117 V AC

MAG2

Test Metals for Ferromagnetism

We've all noticed that permanent magnets attract, and "stick to," certain metals. Such metals are known as *ferromagnetic* materials. In this experiment, you'll test various metallic objects to see whether or not they're ferromagnetic. You'll need some American coins, some silverware, a cast-iron skillet (if you happen to own one), and some canned goods. You'll also need the galvanized and copper electrodes that you used for some of the experiments in the DC section.

How Does a Magnet "Work"?

Whenever the atoms in a piece of ferromagnetic material "line up" to some extent, a *magnetic field* exists around the object. A magnetic field can also be produced by the flow of electric charge carriers in a wire, as you saw when you built the DC galvanometer. In ordinary wires, the charge carriers are electrons. However, any moving charged particle will produce a magnetic field. High-speed protons, atomic nuclei, or *ions* (electrically charged atoms) will do it. So will moving *holes* (electron vacancies within individual atoms).

Magnetic fields produce noticeable and measurable force on ferromagnetic materials. The existence of this force explains why a permanent magnet "sticks" to a steel refrigerator door or a cast-iron pipe, for example. Magnetic force can also produce other effects, such as deflection of an electron beam. That's the principle by which an old-fashioned television picture tube works. Magnetic fields, produced by coils in the picture tube, cause an electron beam to sweep down the screen in a rapid-motion pattern called the *raster*. The image is produced as the electron beam strikes the phosphor on the inside of the screen, causing the phosphor to glow.

The magnetic field around a permanent magnet arises from the same cause as the field around a wire that carries an electric current. The responsible factor in either case is the motion of charge carriers. In a wire, the electrons move along the conductor, being passed among the atomic nuclei in one direction or the other. In a permanent magnet, the "orbital" motion of electrons around the atomic nuclei produces an *effective current* within individual atoms. Each atom in the object becomes a tiny magnet, and because these "nanomagnets" all point in the same direction (more or less), their individual magnetic fields combine to form a larger magnetic field around the entire object.

Permeability

Scientists sometimes portray magnetic fields as *lines of flux*. These lines aren't material things; they're theoretical constructs invented for mathematical convenience. Nevertheless, the orientation of the lines of flux indicates the orientation of the magnetic field, and the number of lines per unit cross-sectional area indicates the intensity of the magnetic field.

In a ferromagnetic material, the flux lines are drawn together more tightly than they are in *free space* (air or a vacuum). As a result, the magnetic field strength in and around a ferromagnetic object increases, compared with how it would be if the object weren't there. *Permeability*, symbolized by the lowercase Greek mu (μ), is a quantitative expression of the extent to which a substance concentrates the lines of flux. By convention, a perfect vacuum has a permeability of 1 ($\mu = 1$). If current is forced through a wire loop or coil in the air, then the flux density in and around the coil is essentially the same as it would be in a perfect vacuum. Therefore, the permeability of air is approximately equal to 1.

If you place a piece of ferromagnetic material such as iron, nickel, or steel inside a coil carrying a large electric current, the number of flux lines per unit cross-sectional area, a factor known as the *magnetic flux density*, can become extremely large. When that happens, you get a magnetic field so strong that you must apply tremendous force to break free from it.

Some specialized metal alloys make the flux density hundreds or even thousands of times greater than it is in free space, so their permeability values range in the hundreds or in the thousands—vastly larger than 1 ($\mu \gg 1$). Some materials don't compress the lines of flux very much; they have permeability values only a little larger than 1 ($\mu > 1$). A few materials can slightly dilate the flux lines compared to their density in free space, so they have permeability values a little bit smaller than 1 ($\mu < 1$). Table MAG2-1 lists some common substances and their approximate permeability values.

Table MAG2-1   Approximate permeability values for various substances.

Substance	Permeability
Dry wood	Slightly less than 1
Wax	Slightly less than 1
Bismuth	Slightly less than 1
Silver	Slightly less than 1
Vacuum	1
Air	1
Aluminum	Slightly more than 1
Nickel	50–60
Cobalt	60–70
Iron	60–100
Steel	300–600
Specialized ferromagnetic alloys	3000–1,000,000

Test Some Objects!

Now that you know all about magnetism, you're ready to test some common household objects to discover whether or not they contain any ferromagnetic metals. Table MAG2-2 shows what happened when I checked out a few objects around my house. Some of the results surprised me! For example, my butter knives tested positive while my spoons and forks tested negative. How about yours?

If you test nonmetallic objects such as paper, cardboard, porcelain, wood, plastic, rubber, or glass, you should always expect negative results. However, if any such object contains embedded ferromagnetic metal, it will test positive. Save the objects after you test them, because you'll need them again in Experiment MAG4. Once you're done testing the food cans, feel free to open them up and consume the contents. (Don't eat the shellfish if you're allergic to it, no matter how good it smells!) Save the can lids.

Warning! *Don't even think about bringing a magnet near any component, device, or system that's intended or designed for medical use.*

Caution! *Don't place a magnet near a computer, flash drive, removable circuit card, recording tape, or old-fashioned diskette. The magnetic field might corrupt the stored data.*

Table MAG2-2   Behavior of various objects in the proximity of a permanent magnet, along with the likely reasons for that behavior.

Item Tested	Attraction	Probable Conclusion
American penny (coin)	No	Sample does not contain ferromagnetic material
American nickel (coin)	No	Sample does not contain ferromagnetic material
American dime (coin)	No	Sample does not contain ferromagnetic material
American quarter (coin)	No	Sample does not contain ferromagnetic material
Paper plate	No	Sample does not contain ferromagnetic material
Plastic plate	No	Sample does not contain ferromagnetic material
Aluminum foil	No	Sample does not contain ferromagnetic material
Butter knife	Yes	Sample must be made of stainless steel
Salad fork	No	Sample does not contain ferromagnetic material
Teaspoon	No	Sample does not contain ferromagnetic material
Cast-iron skillet	Yes	Sample must actually be made of iron (it's heavy!)
Tuna can	Yes	Sample must be made of plated steel
Sardine can	No	Sample does not contain ferromagnetic material
Shrimp can	Yes	Sample must be made of plated steel
Clam can	Yes	Sample must be made of plated steel
Galvanized pipe clamp	Yes	Sample must be made of zinc-coated steel or iron
"Copper" pipe clamp	Yes	Sample must be made of copper-coated steel or iron

Now Try This!

Stack all five of your disk magnets so that they "stick together," forming a larger cylindrical magnet. Repeat the ferromagnetism tests with the same objects you checked before, using this new "five-in-one" cylindrical magnet. Does the larger magnet seem to attract or "stick to" the ferromagnetic objects with greater force than a single disk magnet? Do any of the objects that failed to noticeably attract a single magnet experience any force near the "five-in-one" magnet?

MAG3

Compass Deflection versus Distance

In this experiment, you'll see how the magnetic field around a permanent magnet affects the behavior of a magnetic compass. You'll need the galvanometer that you built in Experiment DC22, along with five ceramic disk magnets. You'll also need a wooden (not metal) yardstick and a flat, horizontal surface that's far away from metallic objects that could influence the compass readings.

Set Everything Up

Retrieve the galvanometer from your "junk box." Remove the wire coil from around the compass. Find a location at which the compass needle points toward magnetic north, unperturbed by the presence of ferromagnetic objects. (In most places, magnetic north is pretty much the same direction as true geographic north, but in a few locations there's a significant difference.) Make sure that the disk magnets are at least 20 feet (ft) away from the compass as you search for the ideal test site. Don't wear anything that might influence the compass readings, either! Ensure that there's at least 5 ft of flat, vacant, horizontal space in all directions around the compass.

Once you've found a suitable test zone, orient the compass so that its needle points precisely at the 0° azimuth mark (the N) on the scale. Then place the yardstick near the compass so that you can measure distances from the center of the compass. The yardstick should lie flat on the test surface, with its "zero end" adjacent to the compass and on a line connecting the 0° (north) and 180° (south) points. Orient the yardstick along an axis going toward magnetic east and west, so that, as you move out along the yardstick, you travel "magnetically eastward."

Figure MAG3-1 shows how the components should be oriented. Don't bring the magnets into the scene until you've properly arranged all the other items.

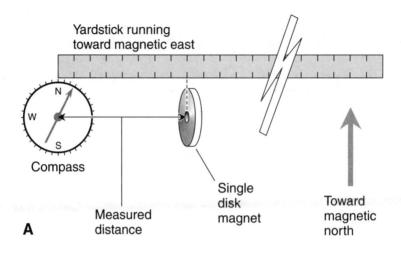

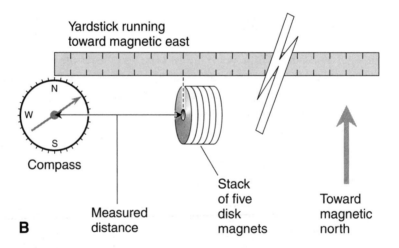

Figure MAG 3-1   *At A, the arrangement for testing the effect of a single disk magnet on a compass. At B, the arrangement for testing the effect of a stack of five disk magnets. Distances should be measured between the center of the compass and center of the nearest magnet face.*

Measure, Tabulate, Graph

Get one of the ceramic disk magnets from the package of five, and keep the other four at least 20 ft away from the test site. Bring the single magnet near the eastern side of the compass as shown at A in Fig. MAG3-1. Turn the disk so that the compass needle deflects toward the east. Carefully orient the disk so that it lies in a vertical plane perpendicular to the yardstick. Place the center of the disk directly east (magnetically speaking) of the compass center.

Move the magnet closer to the compass, always keeping the disk vertical, and always keeping its center exactly on a line going east from the center of the compass. That way, the north and south ends of the compass needle will experience equal but opposite effects from the magnet. The magnet will attract the north end of the needle, while repelling the south end. As you move the disk magnet closer to the compass, the needle deflection will increase. As you move the magnet farther away, the needle deflection will decrease.

Check the needle deflection at distances of 2, 4, 6, 10, 12, 14, 16, 18, and 20 inches (in) from the compass, and then tabulate the results. Table MAG3-1 shows the readings I got. Then plot the data on a coordinate grid, with distance on the horizontal axis and deflection on the vertical axis. My results are shown in Fig. MAG3-2.

Repeat all of the tests with a stack of five disk magnets, placed so that they "stick to each other" to form a single, large, cylindrical magnet. Figure MAG3-1B shows the general arrangement of components. Once again, orient the magnets so that the compass needle deflects toward the east. Measure the deflection with the magnet at distances of 3, 6, 9, 12, 15, 18, 21, 24, 27, 30, 33, and 36 in from the compass center. Tabulate the results (mine appear in Table MAG3-2) and then graph them. I got the points and curve of Fig. MAG3-3.

Compare the results of these tests with the effects of various currents through a wire loop around the compass. You observed and measured these compass deflections when you did Experiment DC22.

Table MAG3-1   Eastward compass deflection versus the distance from a single disk magnet on the east side.

Distance from Compass (inches)	Compass Deflection (degrees)
2	88
4	78
6	50
8	28
10	17
12	7
14	3
16	0
18	0
20	0

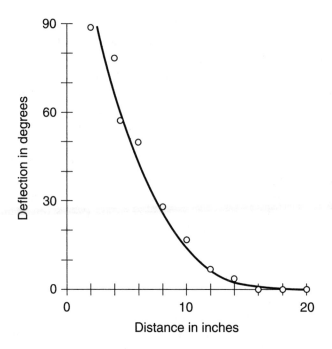

Figure MAG 3-2   *Graph of the eastward compass deflection versus the distance from a single disk magnet. Open circles show measured values; the solid curve approximates the function.*

Table MAG3-2   Eastward compass deflection versus the distance from a stack of five disk magnets on the east side.

Distance from Compass (inches)	Compass Deflection (degrees)
3	89
6	80
9	61
12	41
15	19
18	13
21	6
24	3
27	2
30	1
33	0
36	0

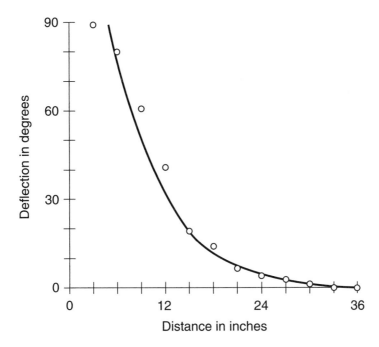

Figure MAG 3-3   *Graph of the eastward compass deflection versus the distance from a stack of five disk magnets. Open circles show measured values; the solid curve approximates the function.*

Now Try This!

Reverse the orientations of the magnets so that the compass needle deflects toward the west, rather than toward the east. When you do this, the magnet will attract the north end of the compass needle and repel the south end of the needle. Compile tables and draw graphs as you did before. How do these "reversed" deflection-vs.-distance functions compare with those of the previous tests?

MAG4

Magnetic Forces through Barriers

In this experiment, you'll see how magnets interact when they're brought near each other, but with barriers between them. You can probably predict that the attraction of opposite poles, and the repulsion of like poles, will take place through nonmetallic obstacles such as paper, just as if the barriers were not there. But what about non-ferromagnetic metal barriers? What about ferromagnetic metal barriers? For this experiment, you'll need everything from Experiment MAG2, along with a can opener.

Feel the Force!

Gather up four disk magnets. Group them into two sets of two. Stack the magnets to obtain two "double disks." Then bring the stacks near each other along a common axis as shown in Fig. MAG4-1. You'll feel the attraction or repulsion. If the stacks attract, they'll come together with a snapping or clapping sound when the distance between them gets less than a certain critical minimum. If the stacks repel, you'll notice how the force increases as you bring them closer together. When the separation distance gets less than a certain minimum, you might have trouble keeping the magnets positioned along a common axis; they'll keep "veering off center."

Test for Attraction

Open up all the cans you tested in Experiment MAG2. Cover the goods inside with aluminum foil, and place the cans in your refrigerator for later use. Wash and dry the lids. Obtain a paper plate, a plastic plate, a sheet of aluminum foil, a cast-iron skillet (if you have one), and all the can lids you just removed. Bring the magnet

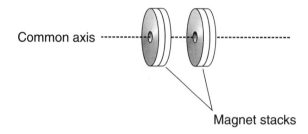

Common axis

Magnet stacks

Figure MAG 4-1 *When you bring two permanent magnets near each other and place them along a common axis, you can feel the magnetic force between them.*

stacks close to each other along a common axis so that they experience mutual attraction. Then have a friend place the plate, the foil, the skillet, and all the different can lids, one at a time, between the magnets (Fig. MAG4-2). How do these obstacles affect the intensity of the magnetic force?

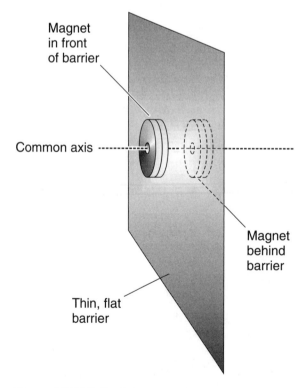

Magnet in front of barrier

Common axis

Magnet behind barrier

Thin, flat barrier

Figure MAG 4-2 *When you place a barrier between a pair of magnets, does the barrier affect the magnetic force?*

Table MAG4-1   Penetration of magnetic force through various objects, as evidenced by attraction or repulsion between permanent magnets. These are my results. Yours might differ, depending on what your cans are made of!

Item Tested	Does Attractive Force Penetrate?	Does Repulsive Force Penetrate?
Paper plate (non-ferromagnetic)	Yes, fully	Yes, fully
Plastic plate (non-ferromagnetic)	Yes, fully	Yes, fully
Aluminum foil (non-ferromagnetic)	Yes, fully	Yes, fully
Cast-iron skillet (ferromagnetic)	Not that I could detect	Not that I could detect
Tuna can lid (ferromagnetic)	Not that I could detect	Yes, but it's reduced
Sardine can lid (non-ferromagnetic)	Yes, fully	Yes, fully
Shrimp can lid (ferromagnetic)	Not that I could detect	Yes, but it's reduced
Clam can lid (ferromagnetic)	Not that I could detect	Yes, but it's reduced

When I tested for magnetic attraction through all of the above-mentioned flat barriers, I got results as shown in the middle column of Table MAG4-1. Non-ferromagnetic metals had no influence on the force of magnetic attraction; I expected that. With ferromagnetic obstacles, I couldn't tell whether or not the mutual attractive force between the magnets increased, stayed the same, decreased, or vanished, because the barrier itself attracted both magnets.

A Little Reminder

It might seem silly for me to repeat this, but I'll say it again anyway: *Don't eat any of the seafood if you suspect that you're allergic to it*, no matter how hungry you get during the course of these tests. If you're not allergic to any seafood (as I am not), you'll think I'm joking. But if you've ever had an allergic reaction to anything (as I have), you'll know I'm serious.

Test for Repulsion

Turn one of the magnets around so that the two "double disks" repel each other when brought into close proximity with nothing in between. Then have your friend insert the barriers, one by one, as you did for the attraction tests.

You should not be surprised to discover that nonmetallic objects such as paper or plastic have no effect on the magnetic force. The repulsive effect operates through these barriers as if there were nothing there at all.

I imagined that non-ferromagnetic metals such as aluminum foil would allow the magnetic force to penetrate fully. My intuition proved correct in these cases, as the right-hand column of Table MAG4-1 shows. However, I thought that ferromagnetic metals would always block the force totally, acting as "magnetic shields." This experiment shattered that long-held misconception!

MAG5

Magnetic Declination

In this experiment, you'll test the accuracy of your hiker's or camper's compass for the purpose of determining true geographical direction. You'll need the magnetic compass from Experiment MAG3. Find a place where you can see the stars on a clear night. Take a small flashlight to help you read the compass.

Declination Defined

The earth's magnetic field, technically called the *geomagnetic field*, has poles just as a bar magnet does. The earth is surrounded by *geomagnetic lines of flux* that converge at the *geomagnetic poles*. However, these convergence zones don't coincide with the *geographic poles*. Therefore, *geomagnetic north* differs from *geographic north* (or true north) at most locations on the earth's surface.

The geomagnetic field interacts with the magnetic field produced by a compass needle, so the needle "tries" to align itself parallel to the geomagnetic lines of flux. The *magnetic declination* at any particular location is the angular difference in a horizontal plane between geomagnetic north and geographic north. (Don't confuse magnetic declination with *astronomical declination*, which is the angle that an object in the sky subtends relative to the *celestial equator*.) In addition to the horizontal force, a compass needle experiences a vertical force called *magnetic inclination*. Most compasses are designed to "ignore" inclination.

In places where a compass needle points toward the east of true north, we define the magnetic declination as a positive angle, expressed in degrees. In locations where a compass needle points west of true north, we define the declination as a negative angle. If we happen to be at a spot where our compass needle points exactly toward true north, the magnetic declination is zero or 0°. At the time of this writing, according to the National Geophysical Data Center (NGDC), the declination was 0° along a line running north and south through the middle of the continental United States.

Locate Polaris

Polaris, also called the *north star* or the *pole star*, is a white star of medium brightness. If you live in the northern hemisphere, Polaris always appears in the sky at the *north celestial pole*. The elevation of Polaris above the northern horizon is equal to your latitude. If you live in Minneapolis, for example, Polaris is 45° above the geographic 0° azimuth point (true north) on the horizon, because Minneapolis is at 45° north latitude.

Polaris rests at the end of the "handle" of the so-called *Little Dipper*. The formal name for that stellar constellation is *Ursa Minor*. A group of stars more familiar to northern-hemisphere observers, the *Big Dipper* or *Ursa Major*, appears overhead on spring evenings, near the northern horizon on autumn evenings, high in the northeastern sky on winter evenings, and high in the northwestern sky on summer evenings. The two stars at the front of the Big Dipper's "scoop," *Dubhe* and *Merak*, are called the *pointer stars* because they "point the way" toward Polaris.

Go outside on a clear night, far enough away from urban lights and smog so that you can see the stars. Find the "scoop" of the Big Dipper. Look outward from the open end of the "scoop," let your gaze wander five or six times the distance between Dubhe and Merak, and you'll find Polaris as shown in Fig. MAG5-1. To double-check, be sure that the star you have found is at the end of the Little Dipper's "handle."

Measure Your Declination

Aim the 0° azimuth marker on your compass scale toward true north, which is the spot on the horizon directly under Polaris. Measure the number of degrees east or west of the 0° marker toward which the compass needle points. Make measurements in several locations so you're sure that nothing (such as underground iron ore) interferes with the readings. If a particular reading is radically different from all the others, disregard the errant reading. When I did this test, my compass needle consistently pointed 13° east of true north, as shown in Fig. MAG5-2. Your results will likely be different, depending on where you live.

Naturally, I wondered about the accuracy of my measurement. I found a "magnetic declination computer" program on the Web site of the NGDC at http://www.ngdc.noaa.gov.

At the time of this writing, the program could be launched by clicking on "Geomagnetic Data & Models" and then selecting "Declination." By the time you read this, the Web site addresses might have changed. (We're dealing with the Internet here, and nothing stays the same on the Internet for long.) In any case, if you enter the phrase "magnetic declination" into your favorite search engine, you should be able to find a site where you can look up or compute the declination at your location.

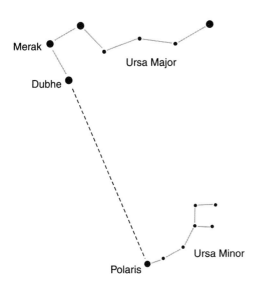

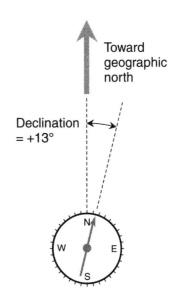

Figure MAG 5-1 *If you can find Ursa Major and Ursa Minor, you can locate Polaris. The orientation of the constellations depends on the time of year and the time of night, but the relative positions never change.*

Figure MAG 5-2 *Orientation of a compass for measuring magnetic declination in the northern hemisphere. This illustration shows the reading I got, indicating +13° declination.*

In the Black Hills of South Dakota where I live, the NGDC site indicated a magnetic declination of about +8°. My reading was therefore 5° off. This puzzled me. I thought I could do better than that, but I took measurements in several places around the neighborhood, with the same result every time.

If You Live "Down Under"

If you're in the southern hemisphere, you can't use Polaris to determine geographic directions because that star is always below the horizon. However, you can sometimes (but not always) locate geographic south using constellations in the southern sky.

As you stand facing toward the south on a clear night, you might see, high in the sky, a group of four stars forming a kite-like shape. This constellation is *Crux*, more commonly called the *Southern Cross*. Just below it, somewhat dimmer, you'll find a star group shaped somewhat like a ladle. That's the constellation *Musca*. Observe Crux and Musca carefully, and make educated guesses as to their centers as shown in Fig. MAG5-3. The center of Crux is easy to decide on, but the

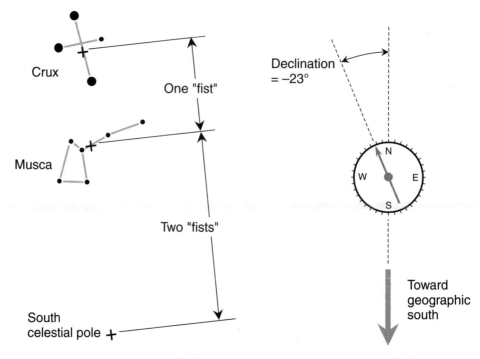

Figure MAG 5-3 *If you live in the southern hemisphere, you can find the south celestial pole by observing the constellations in the southern sky.*

Figure MAG 5-4 *Orientation of a compass for measuring magnetic declination in the southern hemisphere. This illustration shows –23° declination.*

center of Musca is a little more difficult to figure out. Choose a point on the "handle" of the "ladle," just above the "scoop." The two constellation-center points are separated by about 10° of arc, which is approximately the width of your fist at arm's length. Direct your gaze 20° (two "fists") past the center point of Musca on the "non-Crux" side. That will give you a point close to the *south celestial pole.* You can locate *geographic south* (or *true south*) by sighting along the horizon directly below the south celestial pole.

Aim the 180° azimuth marker of your compass scale at a point on the horizon directly under the south celestial pole. Then, just as you would in the northern hemisphere, measure the number of degrees east or west of the 0° point toward which the north end of the compass needle points. If needle points east of north, your magnetic declination is positive; if the needle points west of north, your declination is negative. Figure MAG5-4 shows a hypothetical situation where the declination is –23°.

"Magnetize" a Copper Wire

Permanent magnets don't normally interact with non-ferromagnetic metals. The jumpers that you used in many of the DC and AC experiments consist of copper wire, which is non-ferromagnetic. If you bring a disk magnet near the middle of one of these jumpers, you won't observe any force between the wire and the magnet. But if you drive an electric current through the wire, things change! For this experiment, you'll need the breadboard from the DC experiments, four fresh AA cells, three jumper wires, and five disk magnets.

Set Things Up

Install the AA cells in the breadboard's four-cell holder to obtain a battery that provides 6 volts (V) DC. Connect a jumper between terminals A-3 and A-7. Connect one end of a second jumper to terminal K-3, and leave the other end free. Connect a third jumper between terminals A-3 and K-3, and then physically manipulate the wire so that its central portion hovers about $^1/_2$ inch (in) [13 millimeters (mm)] above the table on which the breadboard sits.

Stack five disk magnets to make a cylindrical magnet. Lay it on the table sideways (so it rolls if you push it). Position the magnet so that the center of one of its faces is approximately $^1/_4$ in (6 or 7 mm) away from the middle of the jumper between A-3 and K-3. When you have everything set up properly, you should have the arrangement shown in Fig. MAG6-1.

Observe the Force

The jumper with the free end can serve as an on/off switch for current through the jumpers. If you touch the free end to terminal D-7 for a moment, you'll short out the battery, thereby delivering a large current through the wires. Don't leave the jumper from K-3 connected to D-7 for more than a couple of seconds at a time.

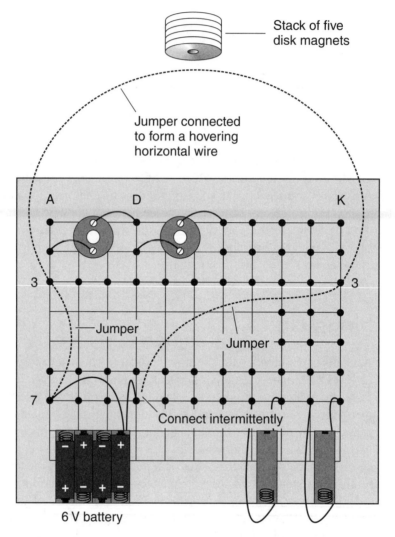

Figure MAG6-1 *Arrangement for observing the effect of magnetic force on a current-carrying wire.*

When you bring the end of the jumper into contact with terminal D-7, watch the hovering jumper in the vicinity of the magnet. The instant that you apply current, the wire will rise up or drop down. When you release the jumper, its center will return to its original position.

Note that the wire does not move closer to, or farther away from, the magnet face. The magnet and the wire do not experience mutual attraction or mutual repulsion. Instead, the force between the magnet and the wire acts at a right angle to the magnet's axis.

Reverse the Magnet

Turn the cylindrical magnet around 180°, so that the face that was farther from the wire becomes the closer face, and the face that was closer to the wire becomes the more distant face. Drive some current through the jumper for about 2 seconds. If the hovering wire near the magnet rose up in the previous test, it will drop down this time. If it dropped down before, it will rise up this time.

Now Try This!

Disconnect both jumpers from the battery. Then reconnect one of them between terminals K-3 and A-7. Connect the second jumper to terminal A-3, and leave the opposite end free as shown in Fig. MAG6-2. Don't change anything else about the

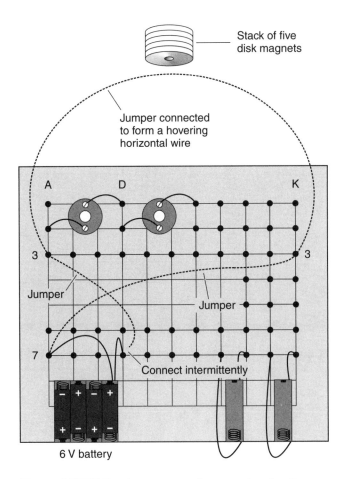

Figure MAG6-2   *Arrangement for reversing the direction of the current through the wires.*

circuit. Don't move the magnet from its position in the second part of the previous experiment.

For a second or two, touch the free end of the jumper from terminal A-3 to terminal D-7. When you do this, you'll send an electric current through the wire in the opposite direction from the way it flowed in the previous two tests. Does this action affect the way the center of the hovering jumper between terminals A-3 and K-3 moves relative to the magnet?

MAG7

Ampere's Law with Straight Wire

For this experiment, you'll need a fresh lantern battery, your magnetic compass, one of the rolls of 22-gauge hookup wire that you bought for the DC experiments, and a couple of lightweight chairs that you can move around easily, such as folding chairs or lawn chairs. You'll also need a pair of safety glasses.

What's Ampere's Law?

Think about what happens when you connect a load to a battery, causing DC to flow through a wire. Electrons "emerge" from the negative terminal of the battery, pass from atom to atom along the wire and through the load, and "enter" the positive terminal of the battery. That concept seems simple, but physicists define electric current in the opposite manner.

The pioneer experimenters in electricity didn't know about atoms and electrons, but they knew that electricity has two opposing "aspects" or "poles," which they called *positive* and *negative* (or *plus* and *minus*). These scientists defined electric current as the movement of charge carriers from positive to negative. Nowadays we know that the electrons, the true charge carriers in a wire, actually move from negative to positive. Nevertheless, the original convention never died away. Scientists still define *conventional current* (also called *theoretical current*) as going from plus to minus.

Imagine a straight wire running toward and away from you, so that you look right down the wire's axis. Suppose that the conventional current flows toward you in the wire. In this situation, the part of the wire close to you has a negative electric charge relative to the more distant part of the wire. According to a principle called *Ampere's law*, magnetic lines of flux take the form of concentric circles around the wire, and the magnetic field "flows" in a counterclockwise direction as shown in Fig. MAG7-1.

Magnetic flux flows in counterclockwise circles

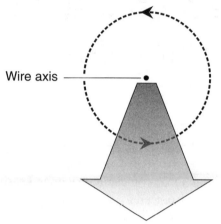

Wire axis

Conventional current flows straight toward you

Figure MAG7-1   *Ampere's law defines the direction in which magnetic flux (dashed circle) flows
around a straight wire that carries DC. Imagine that
the wire runs perpendicular to the page.*

If you know the direction in which the conventional current flows in a wire, then you can determine the direction of magnetic flux flow using a principle called *Fleming's right-hand rule*. Hold your right hand with the thumb pointing out straight (as if you're "hitch-hiking") and the fingers curled. Aim your thumb in the direction of the conventional current traveling in a straight-line path. When you do this, your fingers will curl in the direction of the magnetic flux. Similarly, if you orient your hand so that your fingers curl in the direction of the magnetic flux and then straighten out your thumb, your thumb will point in the direction of the conventional current. If you want to determine the direction of magnetic flux flow relative to the flow of electrons (that is, from minus to plus), use your left hand instead of your right hand.

Where Are the "Poles"?

In basic electricity and magnetism courses, most of us learn that magnetic fields always have well-defined north and south poles, like the poles of a bar magnet or the poles of the cylindrical magnets you've used in earlier experiments. "Conventional wisdom" holds that a *magnetic monopole* can't exist. That means we'll

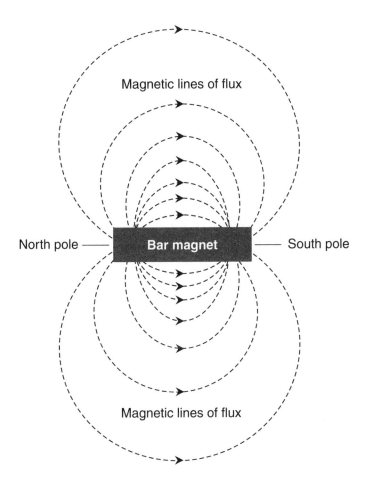

Figure MAG7-2   *Magnetic flux (dashed curves) surrounding a bar magnet diverge from the north pole and converge on the south pole.*

never find a north or south magnetic pole without a "mate" of the opposite polarity somewhere else. A pair of opposite magnetic poles, connected by magnetic lines of flux, constitutes a *magnetic dipole*. Figure MAG7-2 illustrates the positions of the poles, and the shapes of the flux lines (which are actually curves), around a bar magnet. The flux lines have end points at or near the physical ends of the magnet. The magnetic field "emerges" from the north pole, "travels" along the flux lines, and "enters" the south pole.

The flux lines around a current-carrying wire (Fig. MAG7-1) have no apparent end points, but we can indirectly define magnetic poles in this situation.

Imagine two points on one of the flux circles near a current-carrying wire as shown in Fig. MAG7-3. A magnetic dipole exists because of the magnetic-field flow along an arc of the flux circle connecting the points. Let's call this pair of poles a *virtual magnet*. This imaginary object isn't a tangible physical thing like a bar magnet, but a theoretical line segment connecting two points in space. The magnetic field "travels" from the *virtual north pole* to the *virtual south pole*. In the scenario of Fig. MAG7-3, the conventional current flows from left to right; the electrons move from right to left.

Although the virtual magnets near current-carrying wires aren't material objects that we can pick up and toss around, they produce magnetic forces on nearby permanent magnets. Let's observe a virtual magnet in action.

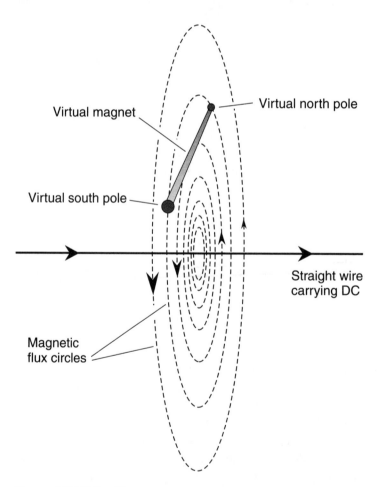

Figure MAG7-3   *We can imagine a virtual magnet near a wire in which constant DC flows.*

Compass Near Wire That Carries DC

Place your magnetic compass on the floor in the middle of a large room with a lot of space, or on an exterior surface such as your yard, patio, or driveway. Turn the compass so that the needle points along the north-south axis.

Place a chair several feet north of the compass. Place another chair a few feet south of the compass. Remove all the wire from a spool of 22-gauge wire, and then straighten the wire out. Strip 1 inch (in) [2.5 centimeters (cm)] of insulation from both ends of the wire. Then string the wire up using the chair legs as "utility poles," stretching the wire tight and straight so that it passes in a north-south direction 1 foot (ft) (30 cm) above the compass as shown in Fig. MAG7-4A. To secure the wire to a chair

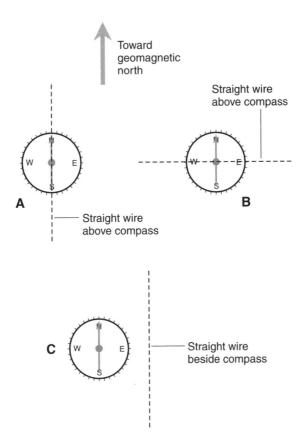

Figure MAG7-4   *Relative positions of straight wire and compass. At A, wire running north-south above compass. At B, wire running east-west above compass. At C, wire running north-south beside compass.*

leg, you can wrap the wire around the leg two or three times, or you can tie the wire in a "half-hitch" knot around the leg. (Don't use any sort of "sticky tape" unless you want to ruin the chair legs!) Bring the ends of the wire around so that they're next to each other, but far away from the compass.

Put on your safety glasses. Touch the ends of the wire to the terminals of the lantern battery for a couple of seconds. The compass needle should rotate. Don't leave the wire connected to the battery for more than 2 seconds at a time.

Reverse the wire terminals, so that the current is forced through the wire in the opposite direction. If the compass needle turned clockwise before, it will turn counterclockwise this time. If it deflected counterclockwise before, it will go clockwise this time.

In both of these situations, the magnetic flux lines around the wire exist as concentric circles in vertical planes that run through the wire at right angles. Near the compass, the flux lines pass in an east-west direction. The resulting magnetic force pulls the compass needle sideways.

Now Try This!

Move the chairs so that they're east and west of the compass, but leave the compass alone. String the wire up tight and straight, passing 1 ft (30 cm) above the compass in an east-west direction as shown in Fig. MAG7-4B. What do you suppose will take place when you connect the ends of the wire to the battery terminals for a second or two? Try it and see.

Set things up as they were in the original experiment. Then push the wire down the chair legs, so that the wire lies on the floor, ground, or driveway. Move the compass 1 ft (30 cm) toward the west (MAG7-4C). Make certain that the compass needle still points straight north and south when no current flows in the wire. What do you think will happen when you touch the wire ends to the battery terminals? Try it and see!

MAG8

Ampere's Law with Wire Loop

You've seen how Ampere's law defines the lines of flux around a straight wire. This principle can also define the flux lines around and through a wire loop. In this experiment, you'll observe the effects of magnetic flux in the center of a DC-carrying wire loop. You'll need a lantern battery, a magnetic compass, two jumpers, a 4 foot (ft) length of AWG No. 8 soft-drawn copper wire, and two pairs of safety glasses.

Old Rule, New Twist

Imagine a circular loop of wire whose axis runs straight toward and away from you. Suppose that you connect a battery to the loop so the conventional current flows counterclockwise (Fig. MAG8-1). Near the wire, the flux lines appear as circles around points on the loop. Farther from the wire, the magnetic field takes a more complicated shape, similar to the field around a bar magnet.

If you know the direction of the conventional current around a circular loop, you can use the right-hand rule to determine the direction of the magnetic flux in the center of the loop. Position your right hand so the thumb points out straight (as if you're "hitch-hiking") and the fingers curl in the direction of the conventional current around the loop. In this situation, your thumb will point in the direction of the magnetic flux at the loop's center.

If you want to determine the direction of magnetic flux flow with respect to the movement of electrons around the loop as portrayed in Fig. MAG8-1, use your left hand instead of your right hand. Remember that the electrons travel from the negative battery terminal to the positive. In this illustration, electrons move from atom to atom around the loop clockwise, contrary to the conventional current.

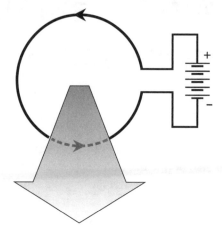

Conventional current flows counterclockwise

Magnetic flux flows straight toward you

Figure MAG8-1   *Ampere's law as it applies to a wire loop (solid circle) that carries DC.*

Another Virtual Magnet

The flux lines around a current-carrying loop have no beginning or end points. Nevertheless, we can define a virtual magnet near the loop's center as shown in Fig. MAG8-2. A virtual magnetic dipole results from the magnetic flux that flows along the loop axis. In the situation shown by Fig. MAG8-2, the plane of the loop runs perpendicular to the page. The loop axis corresponds to a straight line in the plane of the page. The conventional current flows in the direction shown by the arrows on the loop (solid curve), while the magnetic flux on the loop axis flows from left to right (dashed line).

Compass at Center of Loop That Carries DC

Bend the 4-ft length of AWG No. 8 copper wire into the shape of a circular loop with the wire ends approximately 2 inches (in) [5 centimeters (cm)] apart. Attach a black jumper to one end of the loop and a red jumper to the other end. Clip the

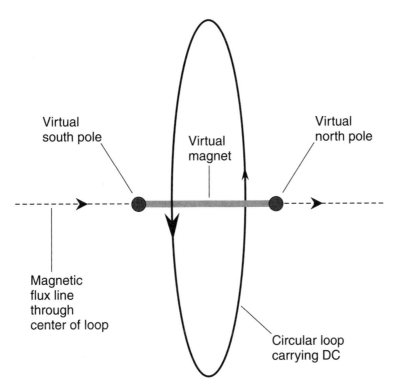

Figure MAG8-2 *A virtual magnet through the center of a wire loop that carries DC.*

"non-loop" end of the black jumper to the negative terminal of your lantern battery. Leave the "non-loop" end of the red jumper free.

Put on your safety glasses. Find a friend, give her another pair of safety glasses, and have her wear them. Have her hold the wire loop in a vertical position so its axis runs horizontally. With one hand, hold the compass horizontally in the center of the loop. Rotate the compass so its needle aligns along the north-south points of the scale. Now tell your friend to orient the loop so its axis runs in a magnetic east-west direction (Fig. MAG8-3A). With your free hand, take the free end of the red jumper and touch it to the positive battery terminal for a moment. The compass needle will turn. Don't leave the red jumper connected to the battery for more than 2 seconds at a time.

Reverse the jumpers at the ends of the loop to drive current through the loop in the opposite direction. If the compass needle rotated clockwise before, it will rotate counterclockwise this time. If it turned toward counterclockwise before, it will now deflect clockwise. In both cases, the magnetic flux through the loop's center follows a straight-line path, going either west to east or east to west and therefore pulling the compass needle sideways.

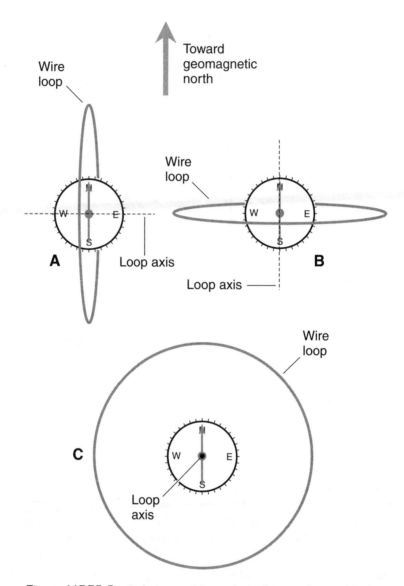

Figure MAG8-3   *Relative positions of wire loop and compass. At A, loop axis running east-west with compass at center. At B, loop axis running north-south with compass at center. At C, loop axis oriented vertically with compass at center. (All views are looking straight down at the compass.)*

Now Try This!

Have your friend rotate the wire loop so that its axis runs horizontally in a magnetic north-south direction (Fig. MAG8-3B). Hold the compass in the center of the loop. Keep the compass aligned with the needle at the "N" (the 0° azimuth point). For a couple of seconds, touch the free end of the red jumper to the positive battery terminal. How does the compass needle behave?

Repeat this experiment with the wire loop oriented in a horizontal plane, so that its axis runs up and down with the compass at the center (Fig. MAG8-3C). Again, touch the free end of the red jumper to the positive battery terminal for a moment. What happens to the compass needle?

Now perform this experiment with the compass outside the loop, always making sure that you hold the compass horizontally with its needle aligned to 0° azimuth when the loop carries no current. Try to predict how the compass needle will move when the loop carries current. As always, treat your battery kindly. Never leave it "shorted out" by the loop for more than 2 seconds at a time.

MAG9

Build a DC
Electromagnet

In this experiment, you'll take advantage of the fact that a current-carrying wire produces a magnetic field by constructing a *DC electromagnet*. You'll need a $^1/_2$-inch (in)-diameter, 12-in-long threaded steel rod, two nuts to fit the rod, 20 feet (ft) of two-wire lamp cord, some electrical tape, a fresh 6-volt (V) lantern battery, your magnetic compass, and your safety glasses.

Theory and Construction

When you place a rod made of ferromagnetic material (called a *magnetic core*) inside a coil of wire and then connect the coil to a source of DC, you get an electromagnet (Fig. MAG9-1). The magnetic flux produced by the current temporarily magnetizes the core. The flux lines concentrate in the core because the core material has high permeability. Therefore, the core acts like a magnet. DC electromagnets have north and south poles, just as permanent magnets do.

The strength of an electromagnet depends on the current in the coil, the number of coil turns, and the permeability of the core. Unless the core has been pre-magnetized, an electromagnet produces a significant magnetic field only when the coil carries current. When the current stops flowing, the magnetic field vanishes almost completely. A small amount of *residual magnetism* might remain in the core after the current ceases, but this field is weak.

Construction

Separate the two-wire lamp cord into identical lengths of single wires. Place one nut at each end of a threaded steel rod. Screw the nuts in approximately 1 in from each end of the rod. Wrap a lamp-cord wire in a neat, tight coil around the rod between the nuts, securing the wire ends to the rod outside the nuts with electrical tape as shown

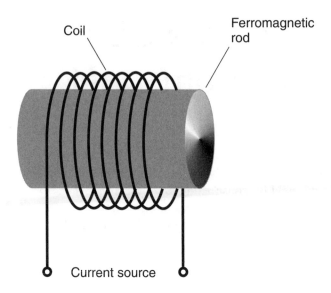

Coil

Ferromagnetic rod

Current source

Figure MAG9-1 *A simple electromagnet consists of a wire coil surrounding a rod made of ferromagnetic material.*

in Fig. MAG9-2. Cut off the excess wire to make each lead 3 ft [1 meter (m)] long. Strip 1 in [2.5 centimeters (cm)] of insulation from each end of the wire.

Warning! *Do not use an automotive battery or other massive electrochemical battery for this experiment. The near-short-circuit produced by an electromagnet can cause the acid from such a battery to boil out and burn you. Clothing offers little or no protection. If the acid gets in your eyes, it can blind you. Use only a conventional 6-V lantern battery. Always wear safety glasses when you work with batteries and high-current devices such as electromagnets.*

Connect one end of the coil wire to the negative battery terminal by twisting the copper strands tightly around the little "spring" at the top of the battery. Leave the other end of the coil free until you want to operate the electromagnet. You can test the device by bringing one end of the rod near objects such as those that you checked for ferromagnetism in Experiment MAG2.

Warning! *Never operate a DC electromagnet near any component, device, or system that's intended or designed for medical use.*

Caution! *Don't place a DC electromagnet near a computer, flash drive, removable circuit card, recording tape, or old-fashioned diskette. The magnetic field might corrupt the stored data.*

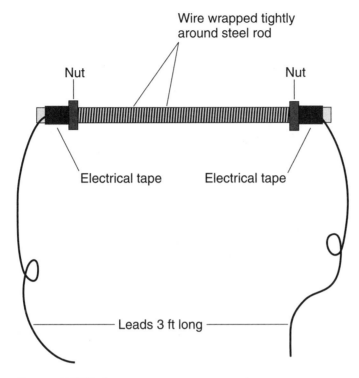

Figure MAG9-2 *Construction of a DC electromagnet using a large threaded rod, two nuts, some insulated wire, and some electrical tape.*

Put on your safety glasses. When you connect both ends of the coil to the battery, the rod will behave as the disk magnets did. Your DC electromagnet will attract objects that contain ferromagnetic material, but won't attract objects that lack ferromagnetic material. Don't leave both ends of the coil connected to the battery for more than 2 seconds at a time.

Which Pole Is Which?

You can use the right-hand rule to determine which end of the rod represents the magnetic north pole. Figure MAG9-3 illustrates the principle. If you look down the rod "endwise" so that the conventional current (from the positive end of the coil to its negative end) appears to rotate counterclockwise as it follows the wire, then the magnetic flux flows toward you, and the near end of the rod constitutes the north pole. If the conventional current appears to rotate clockwise as it follows

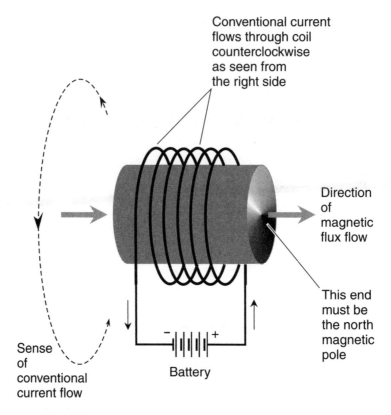

Figure MAG9-3 *The right-hand rule can tell you which end of a DC electromagnet behaves as the north pole.*

the wire, then the magnetic flux flows away from you, and the near end of the rod constitutes the south pole.

Now Try This!

Once you've determined which end of the electromagnet corresponds to magnetic north and which end corresponds to magnetic south, test the device near your compass. To begin, move the electromagnet at least 10 ft away from the compass. Align the compass so that its needle points toward 0° azimuth on the scale.

Bring the magnetic north end of the rod near the compass, keeping the rod in the same plane as the compass face. Does the compass needle rotate even when the electromagnet is disconnected from the battery? If so, you know that the rod holds some residual magnetism as a result of your earlier tests.

Connect both ends of the coil to the battery, leaving the magnetic north pole of the rod near the compass. The electromagnet's north pole will repel the north end of the compass needle and attract its south. Does this result surprise you?

Here's a Factoid!

The north poles of permanent magnets and electromagnets point north because the *geomagnetic* north pole is in fact a *magnetic* south pole! Conversely, the geomagnetic south pole is actually a magnetic north pole. The geomagnetic poles attract and repel the poles of all permanent magnets and all electromagnets—even though the forces act over vast distances.

The term "north pole" arose when early experimenters floated bar magnets on pieces of wood in tubs of water. One end of such a magnet always swung toward the north, so scientists called that end of the magnet its "north pole." Even today, you can find permanent magnets with the poles labeled N and S. Opposite magnetic poles attract, so if you (correctly) imagine the earth as a huge magnet, the S belongs in the Arctic while the N belongs in the Antarctic.

DC Electromagnet near Permanent Magnet

How does a DC electromagnet behave in the proximity of a permanent magnet? In this experiment, you'll find out. You'll need the DC electromagnet that you built in Experiment MAG9, all five of your permanent disk magnets, your 6-volt (V) lantern battery, and your safety glasses.

Bring Them Together with the Electromagnet Off

With the electromagnet powered-down (disconnected from the battery), bring a single disk magnet near one end of the electromagnet's core. When the disk magnet comes within a small fraction of an inch (a millimeter or two) of the end of the rod, the disk should attract and stick to the rod, no matter which face of the disk you test.

Repeat this phase of the experiment with a stack of five disk magnets. You should observe a somewhat stronger force of attraction between the permanent magnet and the end of the rod. The intensity of the force should not depend on which face of the permanent magnet you turn toward the end of the rod.

Bring Them Together with the Electromagnet On

Put on your safety glasses. Do all of the foregoing tests again, but power-up the electromagnet this time. (As always, avoid leaving both ends of the coil continuously connected to the battery for more than 2 seconds.) Bring a single disk magnet near the end of the rod as shown in drawing A of Fig. MAG10-1. Does the

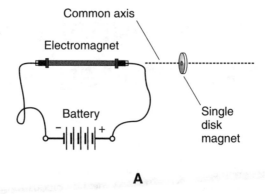

A

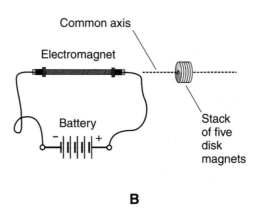

B

Figure MAG10-1   *At A, a single disk magnet near a DC electromagnet. At B, a stack of five disk magnets near a DC electromagnet.*

permanent magnet attract the rod with more force toward one disk face than toward the other face? Do you find that on one disk face the force of attraction is strong, while on the other face the force is minimal or zero? Do you observe a force of repulsion from one disk face and a force of attraction to the other face?

When I did these tests, one face of my disk magnet attracted the end of the core with greater force than it did when the electromagnet received no power. The opposite face of the disk also attracted the core, but with almost no force. A similar effect took place when I stacked up five disk magnets and tested both faces near the end of the electromagnet's core (Fig. MAG10-1B). These results astounded me. I had anticipated a force of attraction toward one disk face and an equal force of repulsion from the other face. Such is the experimenter's life! If things always behaved as expected, we wouldn't have any reason to spend time in the lab, would we?

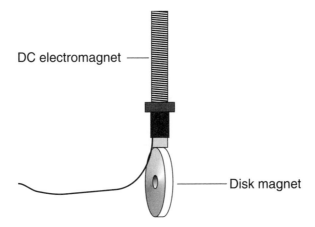

Figure MAG10-2   *Disk magnet "hanging" from the end of an electromagnet's core when no current flows through the coil. What will happen when you connect the electromagnet's coil to a lantern battery?*

Why the Strange Behavior?

Whatever happens when you do these tests, you'll almost certainly discover that when you bring the end of a DC electromagnet near a permanent magnet, the electromagnet behaves differently than another permanent magnet does. Why? Honestly, I'm not sure, but I think it has something to do with the fact that the geometry of the disk magnet differs from the geometry of the electromagnet. Also, the ferromagnetic material in the disk magnets probably differs in permeability from the electromagnet's steel core.

You've seen what happens when you bring identical disk magnets near each other (Experiment MAG4). As you execute the tests in Experiment MAG11, you'll build a second electromagnet, identical to the one used here. You'll connect the two electromagnets in tandem so that they carry equal current when powered-up, bring their ends together, and see how they behave. Would you care to make a prediction now?

Now Try This!

Disconnect the electromagnet from the battery, and orient the electromagnet vertically. Suspend a single disk magnet from the end of the rod as shown in Fig. MAG10-2. Make sure that the disk magnet sticks to the end of the rod, with the curved edge of the disk in contact with the center of the rod's end. What do you think will happen when you power-up the electromagnet? Try it and see! What do you think will take place if you repeat the experiment using the other end of the electromagnet, or if you use the same end of the electromagnet but reverse the battery polarity? Try it and see.

DC Electromagnets near Each Other

In this experiment, you'll see how two DC electromagnets behave when you bring their ends (poles) into close proximity. You'll need the second threaded steel rod that you bought, two nuts to fit the rod, the left over portion of the lamp cord that you ripped apart in Experiment MAG9, some electrical tape, a lantern battery, and your safety glasses.

Build a "Twin" Electromagnet

Wind the left-over single lamp-cord wire around the threaded rod to obtain a new electromagnet. Give the new electromagnet the same number of turns as the one you made in Experiment MAG9, and wind the new electromagnet's coil in the same "sense" as the old one goes.

Here's a trick that you can use to check the "sense" of the coils. Turn one end of your original electromagnet's core so that it faces you. Look directly down along the rod's axis. Follow an imaginary moving point along the wire coil, starting at the near end and moving away. Does the point appear to rotate counterclockwise as it moves down the wire away from you, or does it appear to rotate clockwise? In either case, you should wind your new electromagnet's coil in the same way.

As you did in Experiment MAG9, cut off the excess wire from each end of the new electromagnet's leads. Make each lead 3 feet (ft) [1 meter (m)] long. Strip 1 inch (in) [2.5 centimeters (cm)] of insulation from the end of each lead.

Attraction versus Repulsion

Disconnect both ends of your original electromagnet from the battery. Connect the two electromagnets' coils in parallel. Grasp the exposed copper at one end of each coil (it makes no difference which ends). Twist the copper strands together,

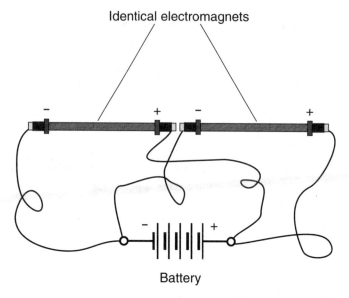

Identical electromagnets

Battery

Figure MAG11-1   *Electromagnets placed end-to-end with opposite poles in proximity.*

obtaining a single, double-thick bunch of copper strands with two wires coming off. Do the same thing with the opposite coil leads.

Put on your safety glasses. Twist one end of the "parallel pair" around the negative battery terminal. Bring the ends of the rods close together so that they come into contact. For a second or two, hold the free end of the "parallel pair" to the positive battery terminal and try to separate the rods. Do rods "want" to stick together? If not, turn one of the electromagnets around and try again; they should stick to each other this time!

When I did these tests, I could feel the rods stick together when opposite poles faced each other as shown in Fig. MAG11-1. But when I brought the like poles together (Fig. MAG11-2), I couldn't notice any mutual repulsion, and I can't explain the reason to my satisfaction. Maybe it has something to do with the fact that the coils don't extend all the way to the ends of the steel rods. Also, the rods have considerable weight. It takes more force to lift them than the magnetic fields can produce between them.

Now Try This ...

Buy three more lantern batteries, identical to the one you already have. Connect all four batteries in parallel by wiring all the positive terminals together and all the negative terminals together using jumpers, as shown in Fig. MAG11-3. You'll get a battery that delivers the same voltage as a single battery, but (in theory) four

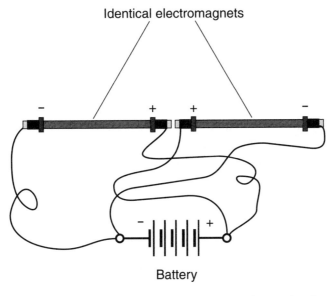

Figure MAG11-2 *Electromagnets placed end-to-end with like poles in proximity.*

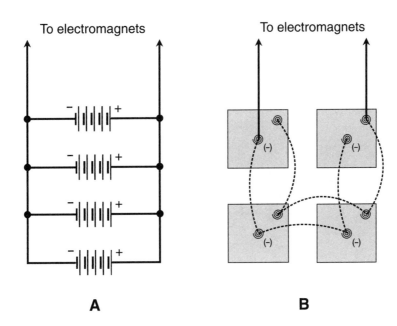

Figure MAG11-3 *Four identical lantern batteries in parallel. At A, the schematic diagram. At B, a pictorial rendition. Dashed curves represent jumpers.*

times the current. Conduct the foregoing tests using the new, more powerful "battery of batteries." Do you notice a greater force of attraction when opposite poles are brought into proximity? Can you feel the force of repulsion when you bring the like poles near each other?

... But Not This!

Do you feel a gnawing desire to find a gigantic battery that can drive huge currents through your electromagnets? I did! *Do not yield to this temptation!* Extreme current can melt the insulation on your electromagnets' wires, burning you or starting a fire. Although I risk coming off like a "worry-wart," I must restate the notice from Experiment MAG9:

> **Warning!** *Do not use an automotive battery or other massive electrochemical battery to provide the power to your electromagnets. Use only conventional 6-volt lantern batteries, and always wear safety glasses.*

MAG12

Build an AC
Electromagnet

You can connect a coil with a ferromagnetic core in series with an electric heater to obtain an AC electromagnet. For this experiment, you'll need one of your existing electromagnets, the series cord that you built in Experiment AC2, an AC-operated heater with two switch-selectable power settings, the 15-ampere (A) power strip that you used in the AC experiments, a roll of electrical tape, and a pair of rubber gloves.

Construction

Before you start setting up the hardware, unplug your series cord to ensure that no current flows through any part of the system. Take the cord apart by removing the electrical tape and unwinding the splices. Set aside the part with the triple outlet on one end and the stripped leads on the other end; you won't need that item for this experiment.

Pick up the cord with the triple outlet on one end, the plug on the other end, and the broken wire in the middle. Twist-splice your electromagnet into that cord, so that the coil is in series with one of the cord conductors as shown in Fig. MAG12-1A. Wear gloves to keep the wire strands from poking your fingers. Wrap the splices individually, and completely, with electrical tape so that *no copper remains exposed*. You should end up with the device diagrammed schematically in Fig. MAG12-1B.

Wear a pair of thick-soled shoes, and keep your rubber gloves on to protect yourself from the danger of electric shock. Switch off the electric heater. Insert the series cord's plug into the power strip. Plug the power strip into a standard 117-volt (V) AC wall outlet. Finally, plug the heater into one of the outlets at the end of the series cord. Your AC electromagnet is now "ready to rumble"!

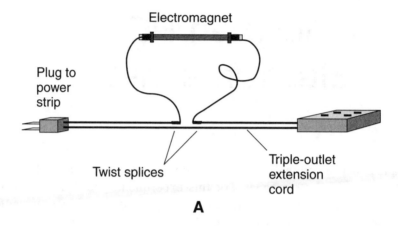

Electromagnet

Plug to
power
strip

Twist splices

Triple-outlet
extension
cord

A

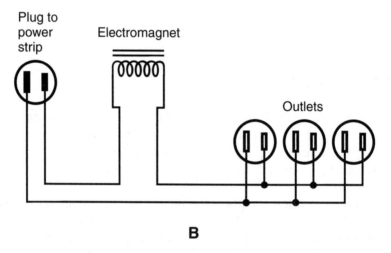

Plug to
power
strip

Electromagnet

Outlets

B

Figure MAG12-1   *Extension-cord modification to connect electromagnet in series with AC outlet. Twist and insulate the splices (drawing A) before you connect the assembly to the power source. Illustration B shows a schematic of the complete system.*

Warning!   *Plugging the coil straight into the power strip without the heater in series would cause a massive current surge through the coil, instantly tripping the power strip's breaker. If the power strip's breaker failed but your luck held, a fuse or a breaker in your household distribution box would "kill" the power. If all the circuit-protection devices failed, your electromagnet's coil insulation would likely melt. A fire could erupt in your household utility wiring. Never connect an electromagnet directly to an AC power source without the heater in series to limit the current.*

Test the Device

Switch the heater on to its lowest power setting. (If your heater has a "fan-only" setting, don't use it; that won't drive enough current through the electromagnet's coils to produce a significant magnetic field.) My heater had two settings, one rated for 750 watts (W) and the other for 1500 W. At a power level of $P = 750$ W, based on a utility voltage of $E = 117$-V RMS, I calculated the RMS current I through the coil as

$$I = P/E$$
$$= 750/117$$
$$= 6.41 \text{ A}$$

When you bring either end of the active AC electromagnet near various objects, the electromagnet will attract things that contain ferromagnetic material, but won't

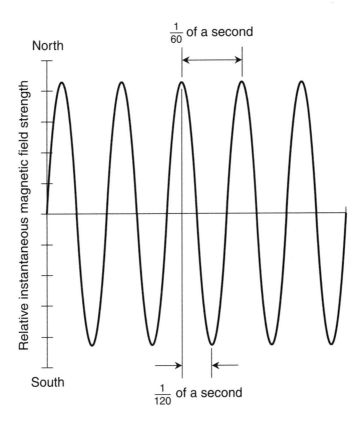

Figure MAG12-2   *Instantaneous field strength as a function of time at one end of an AC electromagnet supplied with 117-V RMS at 60 Hz.*

attract things that lack ferromagnetic material. In these situations, the AC electro-magnet will behave as the disk magnets and the DC electromagnet did—almost!

Now Try This!

Take your other electromagnet, disconnect it from the battery, and bring one end of its core near your AC electromagnet. Slowly draw the ends of the rods together. Do you notice a vibration or "buzz"? You should.

In an electromagnet operating from a 60-hertz (Hz) AC power source, the coil current reverses direction every $1/120$ of a second, the instantaneous current level constantly varies, and the instantaneous magnetic field strength at either end of the device fluctuates as shown in Fig. MAG12-2. Therefore, the magnetic force on any nearby ferromagnetic object pulsates. You can actually "feel the buzz."

When you finish these tests, switch the utility heater off, switch the power strip off, and unplug the entire AC electromagnet assembly from the power strip.

MAG13

AC Electromagnet near Permanent Magnet

In this experiment, you'll bring a permanent magnet into the "flux zone" of an AC electromagnet. You'll need the AC electromagnet that you built in Experiment MAG12, all five of your disk magnets, your compass, and your rubber gloves.

Bring Them Together with the Electromagnet Off

With the AC electromagnet powered-down (the heater switched off), bring a single disk magnet near one end of the steel core as shown in Fig. MAG13-1A. The disk should attract and stick to the rod. When you bring a stack of five disk magnets near the end of the rod (Fig. MAG13-1B), you should feel a stronger sticking force. In every case, you should observe results identical to those that you got when you tested the powered-down DC electromagnet in Experiment MAG10.

Bring Them Together with the Electromagnet On

Put on your gloves. Switch the heater to its low-power setting. When you bring a single disk magnet near either end of the rod, you should observe the same amount of sticking force as you felt when the electromagnet received no current. The same thing should happen with the stack of five disk magnets.

When powered from a standard American household electrical system, the polarity of an AC electromagnet reverses every $1/120$ of a second. Averaged over time, the net magnetic field strength at any point is zero, so the steel core should behave as it does when the coil carries no current. With respect to the sticking

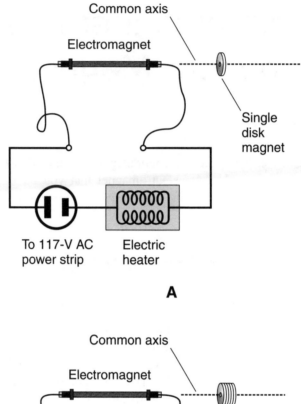

A

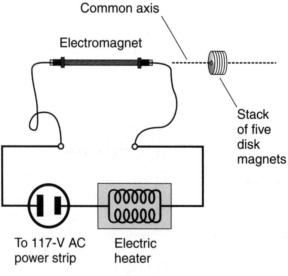

B

Figure MAG13-1   *At A, a single disk magnet near an*
AC electromagnet. At B, a stack of five disk magnets
near an AC electromagnet.

force between the permanent magnets and the rod, I observed precisely that. The *quantitative* effect didn't differ from the situation with the electromagnet powered-down. But the *qualitative* behavior of the disk magnets was strange indeed, particularly when I used a stack of five. With the AC electromagnet carrying current, I could feel the disk magnets vibrate when they came within approximately 1 inch (in) [2.5 centimeters (cm)] of either end of the electromagnet's steel core. Can you feel it as well?

Repeat the above-described tests with the heater switched to its highest power level. Assuming your heater has a maximum power level of $P = 1500$ watts (W) at a voltage of $E = 117$ volt (V) RMS, the current drain I should be

$$I = P / E$$
$$= 1500 / 117$$
$$= 12.8 \text{ A}$$

As before, you'll observe a quantitative sticking force identical to the force with zero current or low current passing through the coil. However, you'll feel a greater vibration effect, particularly with a stack of five disks, when you bring a permanent magnet near either end of the electromagnet's core. Don't leave the electric heater at its high-power setting for more than a couple of minutes.

Bring a Compass near the Electromagnet

For some reason—maybe the experimenter's insatiable curiosity—I tested my compass near the AC electromagnet. With the electromagnet powered-down, I placed the end of its core about $^3/_8$ in (1 cm) from the outer edge of the compass, making sure that the compass needle pointed along the rod's axis as shown in Fig. MAG13-2. Residual magnetism in the rod caused the compass needle to veer from magnetic north, so I had to manipulate the compass and the rod to line them up properly.

I didn't expect the compass needle to move when I activated the electromagnet. I reasoned that, because the AC electromagnet's polarity reverses every $^1/_{120}$ of a second, the forces on the compass needle should average out, and the needle's physical inertia should keep it from "jittering." When I turned on the heater using the low-power setting to energize the electromagnet, the compass needle ignored the alternating magnetic field as expected. When I switched the heater to its high-power setting to maximize the intensity of the alternating magnetic field, the compass needle still did not move.

The surprise came when I switched the heater off to get rid of the alternating magnetic field. The compass needle swung around 180°! Before the application of

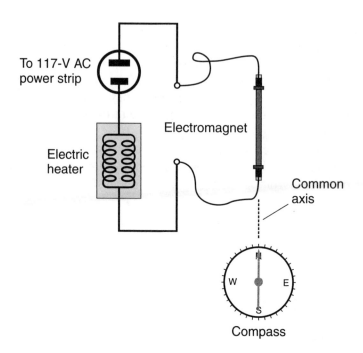

To 117-V AC
power strip

Electromagnet

Electric
heater

Common
axis

W   E

S

Compass

Figure MAG13-2   *Compass near the end of an electromagnet. The rod and the compass needle should lie along a common axis with the electromagnet powered-down. What will
happen when you supply the coil with AC?*

AC, the north end of the needle pointed toward the rod. After the removal of AC,
the south end of the needle pointed toward the rod. Apparently, the "burst" of AC
had reversed the polarity of the residual core magnetism.

When I turned the rod around with the heater off, the compass needle swung
back so that its arrow again pointed toward the end of the rod. Then I applied AC
to the coil once more. The compass needle didn't move. I powered-down the electromagnet. The compass needle still didn't move. I reapplied AC for a few seconds. This time, when I shut the heater back down, the compass needle swung
around 180°! I began to get a little bit obsessed with this weird phenomenon. I
repeated the test several dozen times. On some occasions, the compass needle
reversed direction upon the removal of AC from the electromagnet's core. On
other occasions, the compass needle did not turn around.

I can't explain the behavior of my compass in the vicinity of alternating magnetic
fields—except to suggest that maybe, when an object having residual magnetism
gets "shaken up" by an alternating external magnetic field, the polarity of the residual
field can reverse. When you repeat this exercise many times, do you get results
similar to mine?

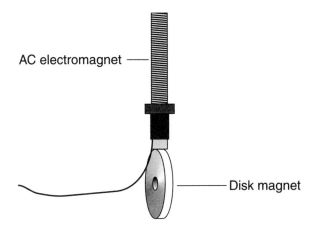

Figure MAG13-3   *Disk magnet "hanging" from the end of a powered-down electromagnet's core. What will happen when you send AC through the coil?*

A Daydream

The evening after I did these tests, I happened to see a television show in which scientists discussed the periodic reversal of the earth's magnetic field, an event that happens every few thousand years. The scientists said that when one of these transpositions occurs, life on earth can be profoundly affected.

I imagined my AC electromagnet's steel rod as a scaled-down imitation of the earth's iron core. I imagined the applied "bursts" of AC as a scaled-down imitation of some massive, sudden magnetic disturbance from the galactic center.

Maybe the earth's core exhibits residual magnetism left over from the solar system's birth. Maybe the polarity of the earth's core reverses when an interstellar magnetic storm comes along and "shakes things up." If that's true, then we can simulate astrophysical events right in our labs!

Now Try This!

Switch off the heater so that the electromagnet receives no current. Orient the electromagnet vertically. Suspend a single disk magnet edgewise from the end of the rod as shown in Fig. MAG13-3. What do you think will happen when you apply power to the electromagnet by switching the heater to its low-power setting? How about the heater's high-power setting? Check it out. When you've satisfied your curiosity, switch the utility heater off, switch the power strip off, and unplug everything from the power strip.

AC Electromagnet near DC Electromagnet

For this experiment, you'll need the AC electromagnet from Experiment MAG13, along with the DC electromagnet that you've used in previous experiments. You'll also need your gloves and glasses.

Bring Them Together with the DC Off

Put on your gloves and glasses. Power-up the AC electromagnet by switching the heater to its low setting. With the DC electromagnet disconnected from the battery, bring one end of its core near one end of the AC electromagnet. Do the ends of the rods attract? When they come into contact, do they stick together?

When I did this test, I felt almost no force of attraction between the ends of the rods, even when the separation between them amounted to only a tiny fraction of an inch [about 1 millimeter (mm)]. When I brought the ends of the rods all the way together, I sensed a slight sticking effect, but far less than I had anticipated.

Repeat the experiment with the heater set for high power, thereby increasing the AC magnetic field strength. Do you notice more force of attraction when you place the ends of the rods in close proximity? Do you notice more sticking force when you bring the rods together and then pull them apart?

Bring Them Together with the DC On

Switch the heater back to low power. With the DC electromagnet still powered-down, hold its "normally negative" end near one end of the AC electromagnet. Then activate the DC for a couple of seconds (Fig. MAG14-1). Do you feel any change in the intensity of the attractive force between the rods? Bring the ends of

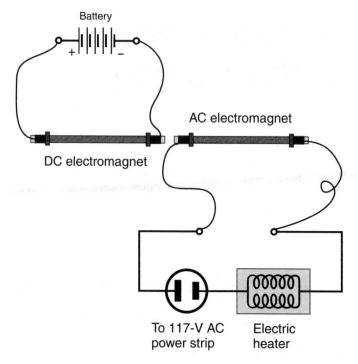

Figure MAG14-1   *An AC electromagnet placed end-to-end with a DC electromagnet.*

the rods together and then slowly draw them apart. How much sticking force do you notice?

When I conducted these tests, I got the same results as I did with the DC electromagnet powered-down. That didn't surprise me. I figured that the alternating attraction and repulsion ought to cancel each other out. I felt the same slight vibration or "buzz" as I'd observed at the conclusion of Experiment MAG12. In fact, with the DC electromagnet powered-up, the "buzz" noticeably increased.

Repeat the experiment with the positive end of the DC electromagnet near one end of the powered-up AC electromagnet (Fig. MAG14-2). Do the two devices behave any differently now, as compared with the case where the negative end of the DC electromagnet was close to one end of the AC core? I didn't expect any difference, and when I did the test, I didn't notice any.

Now Try This!

Perform the previous two tests with the AC electromagnet set for high power. Activate the DC electromagnet for a couple of seconds, deactivate it for a few

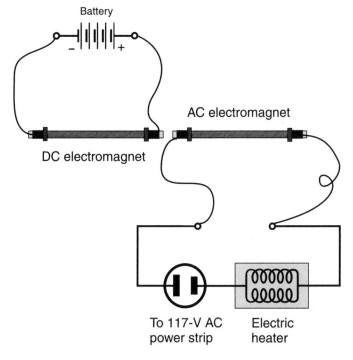

Figure MAG14-2   *Reversing the DC electromagnet's polarity has no effect on its behavior near the AC electromagnet.*

seconds, and then activate it for a couple of seconds again. Do you observe any change in the extent of the attraction or sticking force compared with the low-power AC setting? Do you notice any more vibration or "buzz"?

MAG15

AC Electromagnets near Each Other

In this experiment, you'll test two identical AC electromagnets in close proximity. You'll need both of the electromagnets that you've constructed, along with your trusty old rubber gloves. Before you start this experiment, completely disconnect the DC electromagnet from the battery, and straighten out the stranded copper at the ends of the leads.

Coil Interconnection

Put on your rubber gloves and a pair of thick-soled shoes. Switch off the power strip and the electric heater. Unplug the series cord from the power strip to be "double-sure" that it carries no voltage! Once you've done all that, remove the electrical tape from the connections you made when you built the AC electromagnet in Experiment MAG12.

Wire the demised DC electromagnet in parallel with the existing AC electromagnet as shown in Fig. MAG15-1. Twist-splice the leads from both electromagnets tightly around the copper wires in the series cord. Then wrap the splices thoroughly with electrical tape so that *no copper remains exposed*.

Make certain that the heater is switched off. Then plug the series cord back into the power strip. Be sure that the electromagnets' leads have plenty of slack so they won't snag on anything or get tangled up with each other.

End-to-End Tests

Hold the ends of the rods close together so that they lie along a common axis (Fig. MAG15-1). Switch the heater to its low-power setting. Do you notice attraction or repulsion between the rods? Slowly bring the rods together until they come into contact. Do you feel any vibration? Slowly draw the ends of the rods away from each other. Do they "try" to stick together?

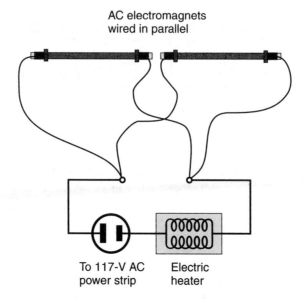

Figure MAG15-1   *Connection of two AC electro-*
magnets in parallel, placed end-to-end. In this
arrangement, the coils share the current equally.

Turn one of the electromagnets around and repeat the entire procedure. You should get a different result.

During the first phase of the foregoing test, I felt no force between the rods, and when I brought the ends into contact, I felt no vibration. When I turned one of the electromagnets around, I detected a slight force of attraction. When I brought the ends all the way together, I felt some vibration. When I slowly separated the ends of the rods, they "weakly tried" to stick to each other.

Switch the electric heater to its high-power setting and go through all of the above-described maneuvers again. Do you notice more repulsion or attraction than you did at low power? Do you feel more vibration or "buzz"? Even at high power, I couldn't detect any repulsive force. I reasoned that it had to exist, but the mass of the rods obscured it. When I turned one of the rods around, I felt considerably more attraction and vibration than I did with the heater set at low power.

Why the Weakness?

Our AC electromagnets produce relatively weak forces because of the action of the current-limiting device (the electric heater). With the electromagnets connected in parallel, they each get half of the total current drawn by the heater, which, as we calculated in Experiment MAG12, is about 6.41 amperes (A) if the heater dissipates 750 watts (W). That means each coil carries only a little more than 3 A.

We must always employ the current limiter to avoid overloading the household utility circuit. Ironically, our AC electromagnets, supplied with 117 volts (V) RMS, exhibit less field strength than our 6 V DC electromagnet does. You might also note that none of our big, heavy electromagnets produce fields anywhere near as strong as those surrounding the little disk magnets from Radio Shack!

Graphical Analysis

The magnetic poles of our AC electromagnets constantly alternate back and forth, completing a full cycle every $^1/_{60}$ of a second and reversing direction every $^1/_{120}$ of a second.

When you bring the ends of the rods together and you *do not* notice mutual attraction, the polarities of the proximate ends are always the same (north-north or south-south), although the instantaneous field strength varies. Figure MAG15-2 graphically illustrates this situation.

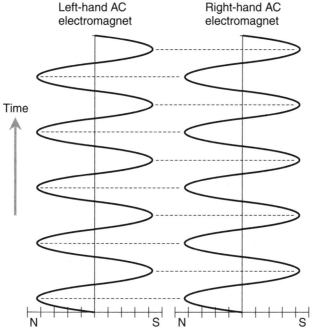

Figure MAG15-2 *Interaction of AC electromagnets as they repel. Each dashed line represents maximum repulsion between like poles in close proximity.*

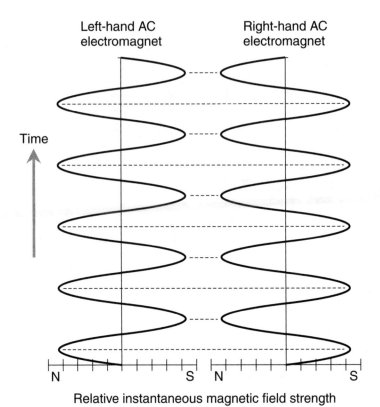

Left-hand AC
electromagnet

Right-hand AC
electromagnet

Time

N   S N   S

Relative instantaneous magnetic field strength

Figure MAG15-3   *Interaction of AC electromagnets as they attract. Each dashed line represents maximum attraction between opposite poles in close proximity.*

When you turn one of the rods around and you *do* notice mutual attraction, the polarities of the proximate ends are always opposite (north-south or south-north), although the instantaneous field strength varies. Graphically, we can portray this scenario as shown in Fig. MAG15-3.

Now Try This!

Keep wearing your safety gloves and shoes. Switch off the power strip and the electric heater. Unplug the series cord from the power strip. Remove the electrical tape from the splices where the electromagnets join the series cord. Take the splices apart so the electromagnets are separated from each other and the cord.

Wire up the electromagnets in series as shown in Fig. MAG15-4. Wrap all three splices with electrical tape, again making sure that no copper remains exposed.

AC electromagnets
wired in series

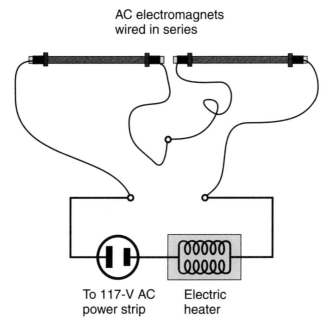

To 117-V AC   Electric
power strip   heater

Figure MAG15-4 *Connection of two AC electromagnets in series, placed end-to-end. In this arrangement, the
full current flows through both coils.*

Plug the series cord back into the power strip, and repeat all of the tests that you
did when you had the electromagnets connected in parallel. How do the forces and
vibrations compare in this arrangement with those in the parallel arrangement?

MAG16

A Handheld Wind Turbine

In this experiment, you'll make an electric fan work as a wind turbine. You'll need one of the rectifier diodes from the DC experiments, your digital multimeter, some electrical tape, and a miniature, single-speed, AC-operated electric fan. You'll also need a motor vehicle, a friend, and a fine day!

Motor versus Generator

In an electric motor operating from standard utility power, 60-hertz (Hz) AC flows through a coil mounted on a rotatable bearing. The alternating magnetic field around the coil interacts with the stable magnetic field between two fixed permanent magnets (Fig. MAG16-1). The attractive force between opposite magnetic poles, the repulsive force between like magnetic poles, and the inertia of the entire assembly causes the coil and its attached shaft to turn at the rate of 60 rotations per second (rps).

An AC generator operates in the same way as an AC motor, but "backward." When a coil rotates inside a stable magnetic field, AC flows in the coil (Fig. MAG16-2). The output voltage depends on the strength of the magnetic field, the number of turns in the coil, and the speed at which the coil rotates. The AC output frequency in hertz is equal to the number of rotations per second that the coil makes. In this experiment, the output frequency and voltage both increase as the wind speed increases.

The "Turbine"

It took me awhile to find the ideal fan for this project. I wanted one designed for operation from standard utility power at 117 volts (V) RMS and 60 Hz. I sought a small object that a person could hold in one hand against a full gale. Eventually I

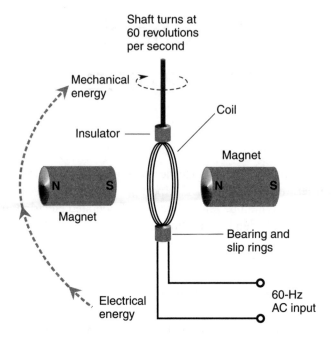

Figure MAG 16-1　*Functional diagram of an AC motor.*

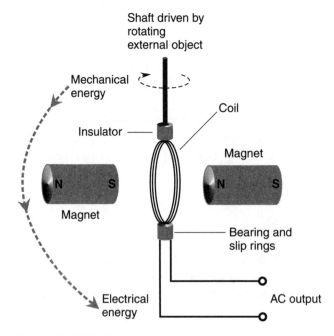

Figure MAG16-2　*Functional diagram of an AC generator.*

found a 4-inch (in) single-speed fan called "Cool-Breeze" at a "Menard's" home supply outlet.

It's important that the fan work from AC, not DC. It's also critical that the fan have only one operating speed: maximum! Any speed control device will interfere with the "backward" operating mode in which you're going to force this thing to function. Your "turbine" must comprise only blades and a motor—nothing else!

The Metering System

Once you've found a suitable fan, cut the plug off the end of its power cord and strip the ends of the leads to expose 1 in [2.5 centimeters (cm)] of bare wire. Connect one of the rectifier diodes left over from your AC experiments in series with the fan and the multimeter as shown in Fig. MAG16-3. The diode's anode should go toward the fan, and the diode's cathode should go toward the meter.

You can "stuff" the diode's cathode lead and the "non-diode" lead from the fan into the meter receptacles. The diode's cathode should go to the positive meter jack, while the "non-diode" fan lead should go to the negative meter jack. Once

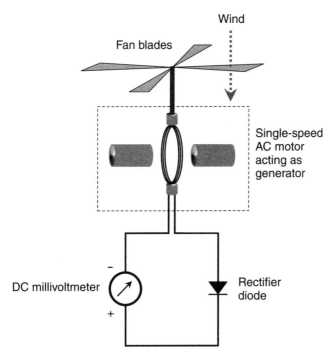

Figure MAG16-3   *Interconnection of fan, rectifier diode, and DC millivoltmeter.*

you've secured the leads this way, use electrical tape to secure the fan cord to your meter so that the leads won't fall out. Finally, switch the meter to show DC milli-volts (mV). My meter had a setting for the range 0 to 2000 mV, which provided the most meaningful readings.

The Test Drive

Now the fun begins! This experiment works best on a windless, dry day. Find a good friend who is willing to hold the fan out the passenger window of your car or truck, even at freeway speeds. Find a road where there is little or no traffic. Drive at various speeds while your friend holds the fan against the wind and records the output voltage as you accelerate.

When you do these tests, you'll experience the temptation to drive at speeds inappropriate for the surrounding traffic. You might want to drive at extreme speeds to simulate a hurricane. Don't engage in any of this dangerous nonsense! Let other drivers build up their anger at someone else's expense. Keep the state troopers happy. And of course, your friend should keep her hands (not to mention your fan) from striking roadside objects!

As your friend compiles a table of millivolts versus miles per hour (such as Table MAG16-1, which shows the results that my friend got), she should try to

Table MAG16-1   Approximate rectified output voltage as a function of the wind speed through a miniature single-speed AC electric fan.

Wind Speed (miles per hour)	Output Voltage (millivolts)
0	0
5	1
10	4
15	17
20	40
25	100
30	150
35	180
40	260
45	330
50	370
55	420
60	450
65	640
70	670

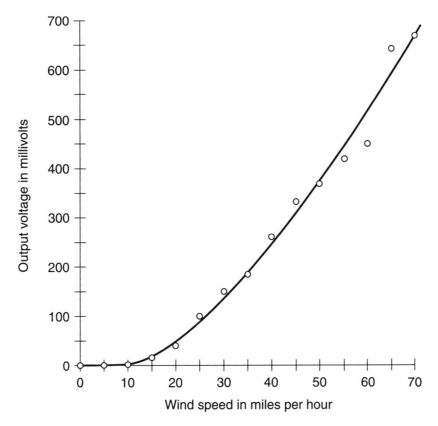

Figure MAG16-4   *Graph of rectified fan output voltage versus wind speed through the fan blades. Open circles show measured values; the solid curve approximates the function.*

keep her hand from interfering with the airflow through the fan blades. My fan had a little table stand attached. This undercarriage served as a handle for keeping a grip on the device without obstructing the gale.

Now Try This!

When you've finished the test drive and have the data table in front of you, plot a graph of output voltage as a function of wind speed. My results appear in Fig. MAG16-4. You'll probably get different numbers, and maybe a different sort of curve. But in any case, you're ready to stand alongside your favorite media daredevil the next time a real hurricane comes your way. The experts can shout out winds speeds in miles per hour. You can chime in with the DC millivolt equivalents.

Alternative Parts Suppliers

All-Electronics
(888) 826-5432
www.allelectronics.com

Design Notes
(800) 957-6867
www.designnotes.com

Electronics Express
(800) 972-2225
www.elexp.com

Jameco Electronics
(800) 831-4242
www.jameco.com

Mouser Electronics
(800) 346-6873
www.mouser.com

Ramsey Electronics
(800) 446-2295
www.ramseyelectronics.com

Suggested Additional Reading

Brindley, K., *Starting Electronics,* 3d ed. Oxford, England: Newnes, 2004.

Cutcher, D., *Electronic Circuits for the Evil Genius*. New York: McGraw-Hill, 2005.

Gerrish, H., *Electricity and Electronics*. Tinley Park, IL: Goodheart-Wilcox Co., 2008.

Gibilisco, S., *Electricity Demystified*. New York: McGraw-Hill, 2005.

Gibilisco, S., *Electronics Demystified*. New York: McGraw-Hill, 2005.

Gibilisco, S., *Teach Yourself Electricity and Electronics,* 4th ed. New York: McGraw-Hill, 2002.

Goodman, B., *How Electronic Things Work—and What to Do When They Don't*. New York: McGraw-Hill, 2003.

Horn, D. T., *Basic Electronics Theory with Projects and Experiments*. New York: McGraw-Hill, 1994.

Morrison, R., *Electricity: A Self-Teaching Guide*. Hoboken, NJ: Wiley Publishing, 2003.

Morrison, R., *Practical Electronics: A Self-Teaching Guide*. Hoboken, NJ: Wiley Publishing, 2003.

Scherz, P., *Practical Electronics for Inventors*. New York: McGraw-Hill, 2000.

Sinclair, I., and Dunton, J., *Practical Electronics Handbook,* 6th ed. Oxford, England: Newnes, 2007.

Index